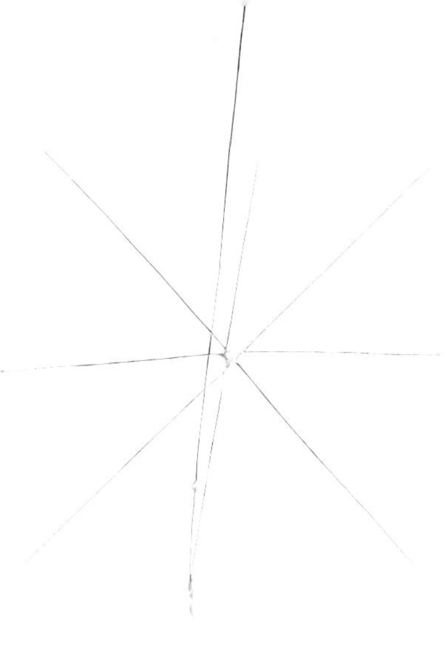

ELLIS HORWOOD BOOKS IN BIOLOGICAL SCIENCES

General Editor: Dr ALAN WISEMAN, Senior Lecturer in Biochemistry, University of Surrey, Guildford

SERIES IN NEUROSCIENCE
Series Editor: Dr A. J. TURNER, University of Leeds

NEUROPHYSIOLOGY: Applications in the Behavioural and Biomedical Sciences
K. A. CLARKE, Department of Physiology, University of Sheffield
NEUROTOXINS IN NEUROCHEMISTRY
Editor: J. O. DOLLY, Reader in Molecular Biology, Imperial College of Science and Technology, University of London
THE NEUROPEPTIDE CHOLECYSTOKININ (CCK): Anatomy and Biochemistry, Receptors, Pharmacology and Physiology
Editors: J. HUGHES, Director, Parke-Davis Research Unit, Cambridge, G. DOCKRAY, The Physiological Laboratory, University of Liverpool, and G. WOODRUFF, Merck Sharp and Dohme, Harlow
PEPTIDE HORMONES AS PROHORMONES
Editor: J. MARTINEZ, Centre CNRS, Inserm de Pharmacologie, Montpellier, France
BRAIN IMAGING: Techniques and Applications
Editors: N. A. SHARIF, Syntex Research, Palo Alto, California, and M. E. LEWIS, Senior Scientist, Cephalon Inc., West Chester, USA

NEUROPHYSIOLOGY
Applications in the Behavioural and Biomedical Sciences

K. A. CLARKE B.Sc., Ph.D.
Department of Biomedical Science
University of Sheffield,

ELLIS HORWOOD LIMITED
Publishers · Chichester

Halsted Press: a division of
JOHN WILEY & SONS
New York · Chichester · Brisbane · Toronto

First published in 1989 by
ELLIS HORWOOD LIMITED
Market Cross House, Cooper Street,
Chichester, West Sussex, PO19 1EB, England
The publisher's colophon is reproduced from James Gillison's drawing of the ancient Market Cross, Chichester.

Distributors:

Australia and New Zealand:
JACARANDA WILEY LIMITED
GPO Box 859, Brisbane, Queensland 4001, Australia

Canada:
JOHN WILEY & SONS CANADA LIMITED
22 Worcester Road, Rexdale, Ontario, Canada

Europe and Africa:
JOHN WILEY & SONS LIMITED
Baffins Lane, Chichester, West Sussex, England

North and South America and the rest of the world:
Halsted Press: a division of
JOHN WILEY & SONS
605 Third Avenue, New York, NY 10158, USA

South-East Asia
JOHN WILEY & SONS (SEA) PTE LIMITED
37 Jalan Pemimpin # 05–04
Block B, Union Industrial Building, Singapore 2057

Indian Subcontinent
WILEY EASTERN LIMITED
4835/24 Ansari Road
Daryaganj, New Delhi 110002, India

British Library Cataloguing in Publication Data
Clarke, K. A.
Neurophysiology.
1. Man. Nervous system. Physiology
I. Title
612'.8

Library of Congress data available

ISBN 0–7458–0567–1 (Ellis Horwood Limited)
ISBN 0–470–21554–2 (Halsted Press)

Typeset in Times by Ellis Horwood Limited
Printed in Great Britain by The Camelot Press, Southampton

Table of contents

Preface

Psychology is a fascinating subject. As a result, the student of psychology or other behavioural science is often pitched into a consideration of psychological processes with little or no knowledge of the neurophysiology which underpins them. In most psychological texts the physiological aspects of brain function are condensed into a short introductory section. On the other hand, most textbooks of neurophysiology are much longer and more detailed than is required for this purpose. The aim of the present text is to provide a brief account of some of the aspects of neurophysiology which might help to produce a firm basis for further study of the brain.

Acknowledgements

I would like to thank Julie Cookson and Andy Parker for their technical drawing of the illustrations. I would also like to thank my wife for the idea for this book.

1

Functional neuroanatomy

Possession of a nervous system is not universal throughout the animal kingdom. It is therefore of interest to identify the qualities of life with which it is associated. Its presence confers on the organism the capacity to sample and process aspects of its internal and external environments as a basis for suitable action. Such action might involve the stimulation of effector cells, examples of which are muscles and glands.

1.1 EVOLUTIONARY DEVELOPMENT OF THE NERVOUS SYSTEM

A number of stages can be identified in the evolution of the nervous system (Fig. 1.1). Simple, unicellular organisms such as the amoeba are in fairly intimate contact with their environment and can move towards or away from a limited range of stimuli. In multicellular organisms, however, cellular specialization occurs. In some primitive multicellular organisms the effector cells also detect the stimulus. For instance, in the sponge, the muscles which control the ocula are also stimulus-sensitive. Higher up the evolutionary tree, some of the epithelial cells overlying the muscle become specialized for irritability — responsiveness to particular stimuli. Here we have the beginnings of a sensory receptor system.

The next important developmental stage is seen, for example, in the tentacles of sea anemones, where the receptor cells give off fine processes specialized for conductivity and these communicate with the effector cells. With the advent of such a communicating system, receptors and effectors may be located in separate, perhaps more appropriate positions. The advent of a neural network interposed between receptor and effector opens up several new possibilities. Not only does it act as a transmission system so that the receptors can be located remotely from the effectors, but it also allows the integration of information from many receptors, permitting more complex processing and plasticity of response. With advancing evolution, larger, elongated animals appeared. This lengthening was often accomplished by replication of basic structural units known as segments. Because of this the nervous system also developed on a segmental basis. While we still retain this basic segmental structure in our spinal cord, there are considerable anatomical and functional links

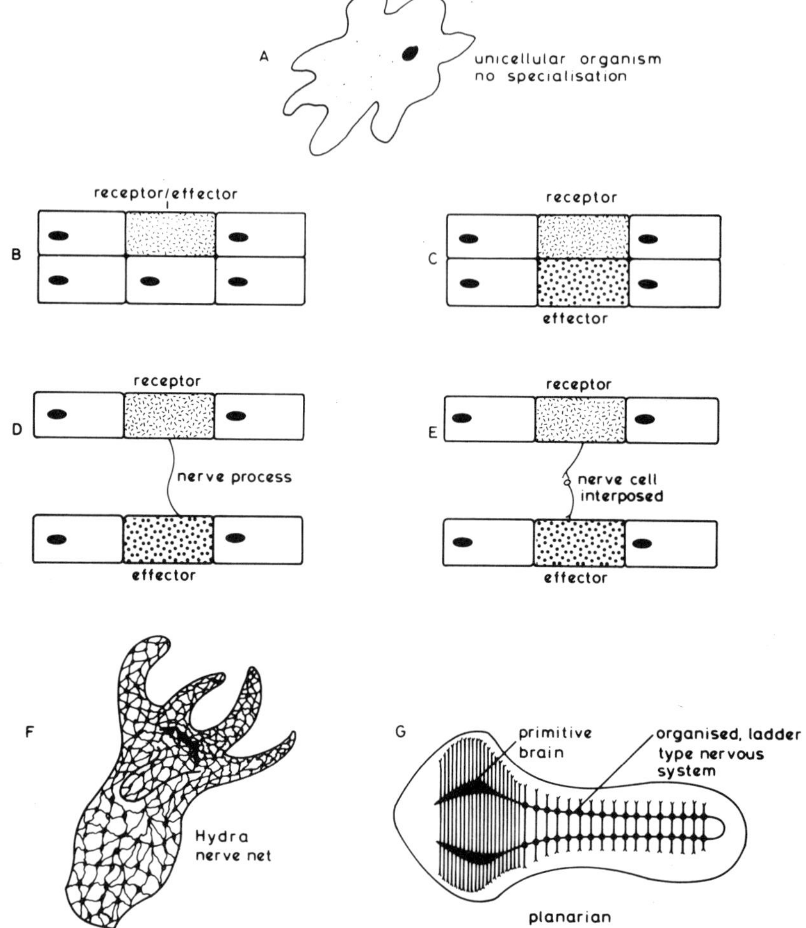

Fig. 1.1 — Some of the major stages in the evolutionary development of the nervous system.

between the segments. By the level of animals with primitive backbones, the sub-epithelial nerve network has become far more complex and has migrated inwards to form the spinal cord. The cell bodies of the receptors have also moved from the surface, leaving a specialized ending connected to the cord by means of a nerve process. Further development has occurred in the form of cephalization — aggregation of neural material at one end of the spinal cord which massively increases processing power.

1.2 STRUCTURE OF THE NERVOUS SYSTEM

1.2.1 Central and peripheral nervous system
At the gross morphological level the nervous system may be divided into central

(spinal cord and above) and peripheral (everything else) components. In the peripheral nervous system (PNS), the sensory receptor is connected to a sensory afferent nerve fibre which conveys the message toward the central nervous system (CNS). This sensory afferent very soon joins a mixed peripheral nerve which contains many other fibres, both afferents and efferents (which convey signals from CNS to effectors or in some cases modulate receptor sensitivity). Much variation is seen in receptor structure, ranging from simple specializations of nerve endings to elaborate collections of cells closely associated with, but distinct from, the nerve ending. Just outside the vertebral column, the bony casing of the spinal cord, the mixed nerve splits into dorsal and ventral roots. The cell bodies of sensory afferents are in the dorsal root ganglia which lie close to the spinal cord.

1.2.2 Spinal cord

We have retained the segmental nature of the spinal cord during our development from lower, segmented animals. Thus a pair of spinal nerves arises from each of the 31 segments of the spinal cord, each nerve innervating a well-defined region of the body. Sensory afferents enter the spinal cord at its dorsal aspect while efferents exit from the ventral surface (Fig. 1.2). Equivalent systems exist for nervous innerva-

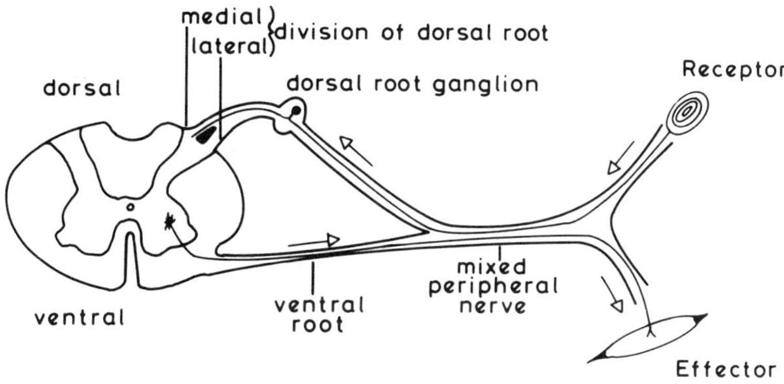

Fig. 1.2 — Distribution of afferent and efferent nerves from a segment of the spinal cord (supply to one side of body shown).

tion of the head by means of the cranial nerves which enter and leave the nervous system via the brainstem which is rostral to, and continuous with, the spinal cord.

The outer portions of a fresh spinal cord cut in transverse section have a glistening white colour due to the presence of myelin, the fatty substance found wrapped around many nerve fibres, indicative of the preponderance of nerve fibres in this region. It is often referred to as white matter. The central core of the cord is butterfly shaped, pinky brown in colour, and divisible into dorsal and ventral horns, the colour reflecting the great number of cell bodies in this region. It is often known as grey matter, not, as one might expect, pinky brown matter. The terminology is derived from the colouration produced by early methods of preservation of neural material.

Once a sensory afferent reaches the spinal cord it may contribute to one or more pathways according to the complexity of processing required. It might provide sensory input to power segmental reflexes. In some instances the sensory input projects directly to the motoneurons in the ventral horn. This direct route is illustrated by afferents from receptors in muscle which detect increases in muscle length and initiate reflex contraction, for instance in the well-known knee-jerk reflex. At a slightly more complex level the afferent may synapse with another neuron, an interneuron, shortly after it enters the dorsal horn of grey matter, and activate a chain of one or more interneurons which eventually synapse with the motoneuron. An example of this would be the spinal reflex for the rapid withdrawal of a limb from a painful stimulus. Short interneurons carry much information across to the opposite side of the cord and between adjacent segments.

Alternatively, the neuron with which the sensory afferent synapses in the dorsal horn may give rise to a long axon which ascends the cord in the white matter to reach higher centres. An example of this would be the ascending systems for pain perception. Finally, the sensory afferent may not synapse in the cord at all, but as soon as it enters via the dorsal root turn and ascend in the white matter to reach higher centres. An example of this would be many of the low-threshold tactile afferents. Of course, any particular afferent entering the cord may branch or supply several interneuronal types so that the same sensory input may be subjected to a number of levels of complexity of processing.

The white matter consists of ascending fibre systems conveying sensory information upwards to higher centres, and a variety of descending fibre systems carrying instructions from higher centres to control motoneuron activity and to adjust the sensitivity of spinal segmental mechanisms (Fig. 1.3). While much of the nervous system is devoted to the control of our skeletal muscles, allowing us to develop a multiplicity of motor skills, another important aspect is the control of our more vegetative side such as digestive and cardiovascular functions. These are the province of the autonomic nervous system, the distribution and functional aspects of which are described in Chapter 10. The descriptive term brain or encephalon is that part of the CNS within the cranial cavity. It is divisible into hind-, mid-, between- and endbrain components. The hind- and midbrain parts are often referred to collectively as the brainstem, while the between- and endbrain portions are called the forebrain. The major subdivisions of the nervous system are shown in Fig. 1.4 and diagrams of inferior, lateral and medial aspects of the brain in Figs 1.5, 1.6 and 1.7 respectively.

1.2.3   Rhombencephalon (hindbrain)

The upper end of the spinal cord is continuous with the lower end of the brainstem, an area known as the medulla oblongata. The brainstem gives rise to a number of cranial nerves, contains neuronal concentrations of importance for respiration, motoneuron control, reticular formation and the autonomic nervous system. Hence the significance of assessing brainstem viability in the estimation of the possibility of functional recovery from head injury. Sitting on the superior surface of the upper

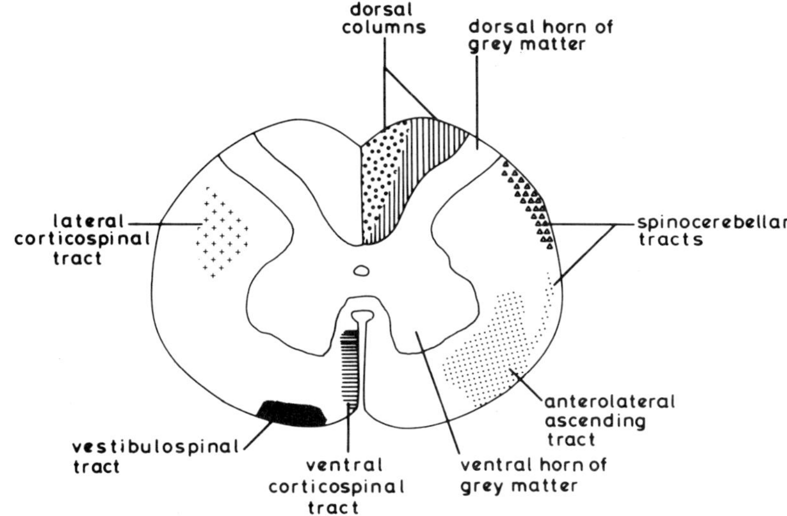

Fig. 1.3 — Transverse section through spinal cord showing location of some major ascending and descending nerve fibre tracts of the white matter.

Spinal cord	Assembly of similar segments	
Brainstem	Hindbrain (rhombencephalon)	medulla oblongata (myelencephalon) cerebellum and pons (metencephalon)
	Midbrain (mesencephalon)	superior and inferior colliculi red nucleus mesencephalic reticular formation substantia nigra
Forebrain (prosencephalon)	Betweenbrain (diencephalon)	thalamus hypothalmus
	Endbrain (telencephalon)	cerebral hemispheres cerebral cortex basal ganglia limbic structures

Fig. 1.4 — Major subdivisions of the mammalian CNS.

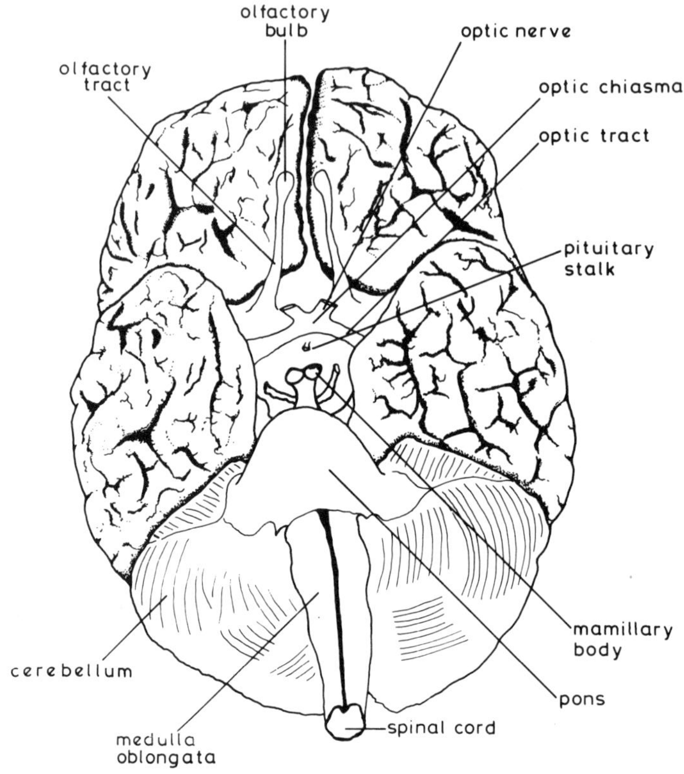

Fig. 1.5 — Inferior surface of brain. Origins of many of the cranial nerves have been omitted for the sake of clarity. Note slight physical asymmetries between left and right sides.

part of the brainstem and attached to it by means of three pairs of legs or peduncles is the cerebellum. In humans this fist-sized piece of tissue is important in the learning, planning and execution of motor skills.

1.2.4 Mesencephalon (midbrain)

The uppermost reaches of the brainstem constitute the midbrain. The midbrain is a relatively small portion of the brainstem. Amongst its contents are four small bodies known as the colliculi, a superior pair important in visual reflexes and an inferior pair significant in auditory reflexes. Also found here are elements of the reticular formation and two areas important in motor control: the red nucleus and the melanin-pigmented substantia nigra. Beyond this uppermost portion of the brainstem is the diencephalon.

1.2.5 Diencephalon (between brain)

This is a collective name for a number of important brain structures, such as the thalamus. The thalamus contains important relay nuclei for the sensory and motor systems, elements of the reticular formation for modulation of the sensitivity and

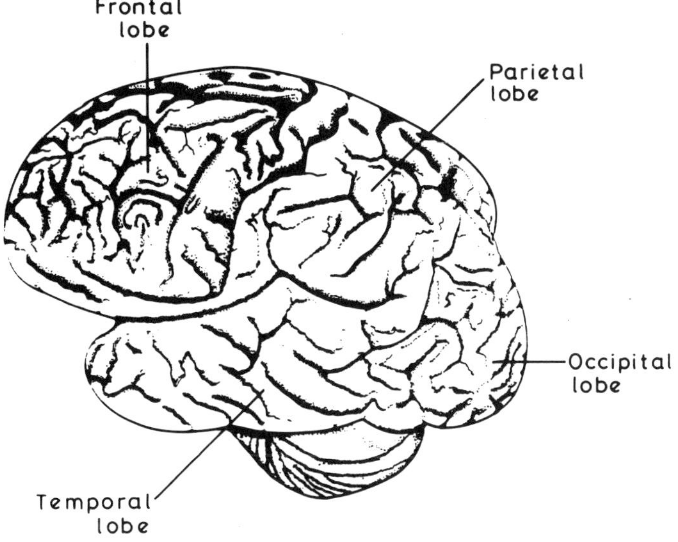

Fig. 1.6 — Lateral view of brain showing major lobes.

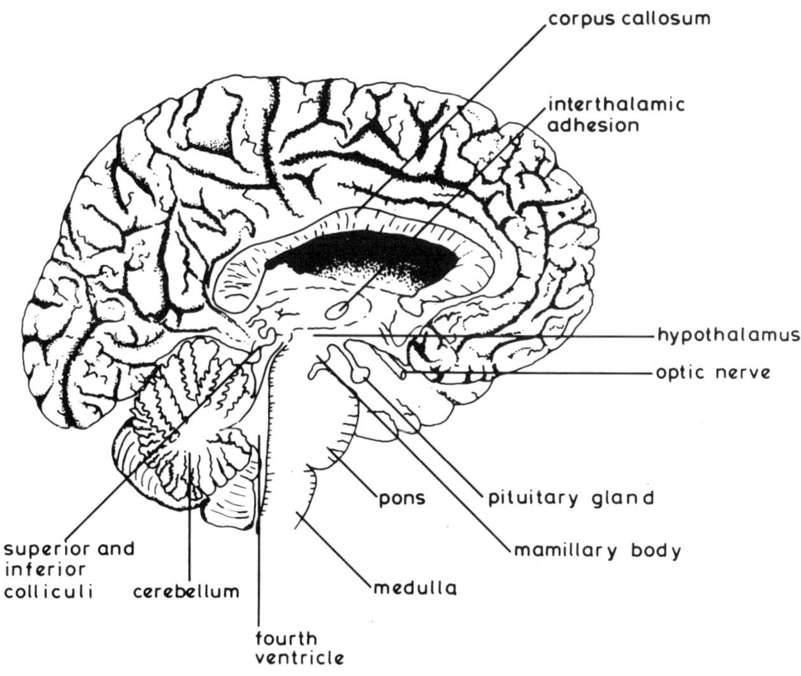

Fig. 1.7 — Midsagittal section through brain.

state of arousal of the brain, and nuclei which project to the cortical association areas of importance in recognition, interpretation and memory. It is thought to be a major site of action of general anaesthetics. Also in the diencephalon is the hypothalamus, an important integrating centre for the autonomic nervous system and various ingestive behaviours, together with the intimately associated pituitary gland over whose functions it exerts powerful control. The pituitary gland in turn controls much of the endocrine system.

1.2.6 Telencephalon (endbrain)

The telencephalon or endbrain is composed of the cerebral hemispheres, some parts of which, particularly in humans, are enormously developed. The two hemispheres are separated by the medial longitudinal fissure, at the bottom of which lies the corpus callosum, a collection of several million transversely running fibres which relay information between equivalent points of the cerebral cortex. One radical treatment for some severe forms of epilepsy is severance of an appropriate segment of the corpus callosum, which prevents the epileptiform activity being transmitted to its equivalent point on the opposite hemisphere. However, such partial separation of the two cerebral hemispheres can cause problems of perception, since the apparent gross anatomical symmetry between the hemispheres belies considerable functional asymmetry.

Covering the surface of each hemisphere to a depth of up to 3 mm is the cerebral cortex, the surface of which is folded into ridges known as gyri, separated by ditches called sulci. This convoluted arrangement greatly increases the cortical surface area available for processing. Indeed there is a good correlation between the extent to which the cortex is furrowed and the position of the species on the evolutionary tree. Some of the many putative functions of the cerebral cortex will be discussed in later chapters.

Buried deep within the cerebrum are a group of nuclei collectively referred to as the basal ganglia. Pathological changes in these nuclei, for instance in Parkinson's disease, have devastating effects upon many aspects of movement. These will be considered in more detail in Chapter 9. One final important component of the telencephalon is the limbic system, a collective term which includes the hippocampus, amygdala, mamillary bodies and septum. These structures are thought to be involved in a number of functions including memory and emotion.

1.3 EMBRYOLOGICAL DEVELOPMENT

We have already considered the evolution of the nervous system. Finally in this introductory chapter we need to take a brief look at the development within the individual, the embryology. There are interesting parallels between evolutionary and embryological development. Indeed, during its growth, the nervous system of the embryo passes through the embryonic stages of its ancestors.

At an early stage in its development the embryo consists of three germ layers known as ectoderm, mesoderm and endoderm (Fig. 1.8). It is from these germ layers that all the various tissues and organs of the body will develop, including the nervous system. The initial stage of neural development, which occurs in the first few weeks of embryonic life is marked by the formation of a trench-like depression among the

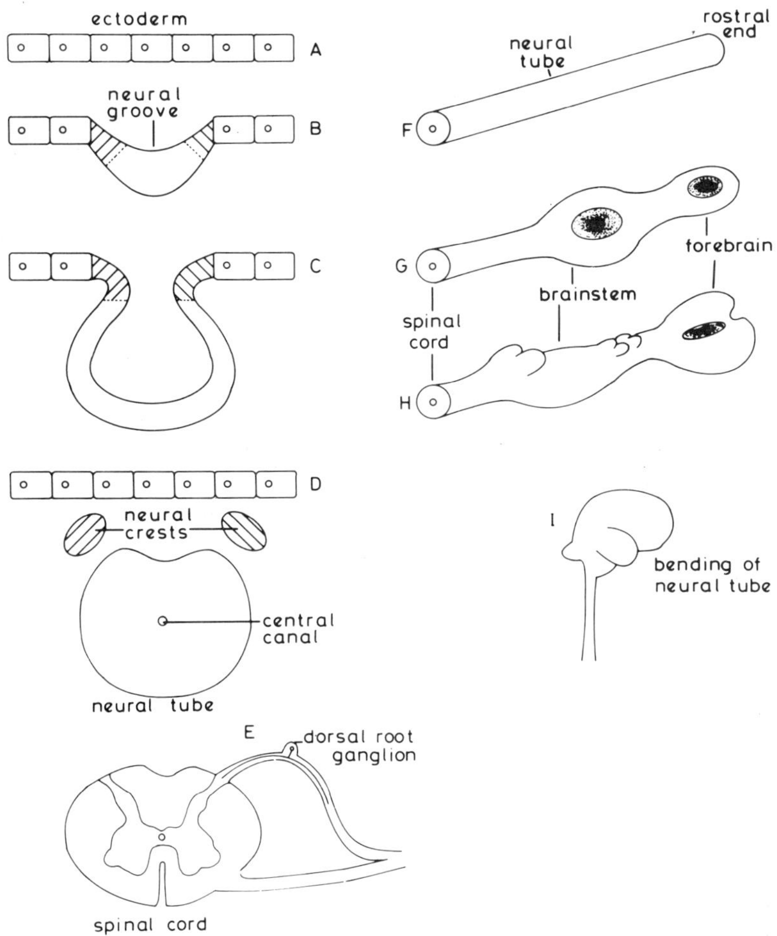

Fig. 1.8 — Embryological development of nervous system from ectodermal germ layer.

cells of the dorsal surface of the ectoderm and running the length of the body. As the channel deepens and widens, the outer parts of its upper edges peel back and detach to form neural crests, while the more medial parts meet and form an enclosed neural tube which then sinks into the middle, mesodermal layer of the embryo. The neuro-ectoderm cells of the neural crests will become the dorsal root ganglia, which contain the cell bodies of the sensory afferents. Each of these cells sprouts two processes; one grows towards and joins up with the dorsolateral aspect of the neural tube, while the other grows in the opposite direction to form a sensory afferent nerve. Some cells in the neural tube develop axons which pass out from the anterolateral aspect of the tube to form the efferent component of the PNS.

Meanwhile, the neural tube differentiates into central grey and surrounding white matter, leaving the centre of the tube open as a fluid-filled spinal canal, which, at two points towards its upper end, widens and opens onto the dorsal surface.

Enormous growth of the surrounding neural tissue will cause these initially surface depressions to lie deep within the brain, forming a series of buried, interconnected fluid-filled cavities known as ventricles. The neural tissue around the more caudal of the two depressions differentiates to form the brainstem, that between them develops into the midbrain, and that around the upper depression forms the forebrain. The most obvious feature of the human forebrain is the enormous development of the cerebral hemispheres. The neural tube does not remain straight during all these changes. Bends occur above and below the developing midbrain, complicating the final appearance of the whole structure.

1.4 NEURONAL ELEMENTS

The vertebrate nervous system contains mainly two sorts of nerve cell. One type is concerned with the transmission and integration of information, while the other sort, glial cells, perform important supportive and metabolic roles. Information-carrying nerve cells consist of a cell body or soma from which arises, at a specialized region called the axon hillock, a single long process known as the axon, which carries information away from the soma, plus a number of shorter processes called dendrites which themselves bear secondary and tertiary dendrites and convey information towards the soma (Fig. 1.9). The extent to which the dendritic tree develops and is

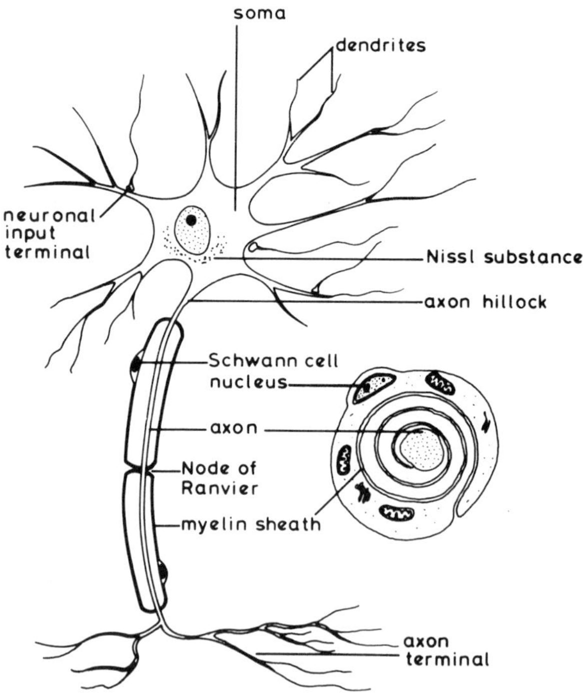

Fig. 1.9 — Appearance of a 'typical' neuron (left). There is, however, enormous variation in neuronal appearance. A neuron with a myelinated axon has been chosen so that its relationship with the glial cells which supply the myelin (in this case Schwann cells) can be shown (cross-sectional view of the axon on the right).

maintained is thought to depend on the quantity and nature of information which passes into it.

Glial cells envelop the axon over much of its length. In some cases these glial cells produce myelin, the presence of which significantly increases the conduction velocity of the nerve. In the PNS these glial cells are known as Schwann cells, while their equivalents in the CNS are called oligodendrocytes.

Small gaps in the myelin covering occur between adjacent glial cells. These gaps are called nodes of Ranvier and are important in information transmission. In non-myelinated axons the association between the glial cell and the axon is much less intimate and the fatty substance myelin is not formed. The significance of the presence of myelin for axonal function will be considered in more detail in Chapter 2. At its termination the axon breaks up into fine branches called telodendria. Inflow of sensory information to the nervous system, integration within the CNS, and subsequent outflow of motor or other information from the CNS is by means of chains of neurons grouped together in particular ways. A gap called a synapse physically separates one neuron from preceding and succeeding ones. A chemical neurotransmitter is often used to carry the message across this gap. Within this complex system one sees specialization of neuronal elements to perform the various necessary functions. In particular we need sensory receptors to detect environmental changes, a transmission system to convey information back and forth, and an integrating system to decide the best response to any stimulus. Let us consider the first link in the chain — the sensory receptor.

1.4.1 Sensory Receptors

The sensory receptor system is used to sample the environment at as many points as possible as the basis for a suitable response. In addition to the universe outside our bodies from which we need to extract information there is a complex environment inside our bodies. Since humans are homeostatic animals, that is we regulate many aspects of our internal environment, the ability to detect and rectify any changes is of major importance.

The total amount of sensory bombardment hitting the CNS is enormous and considerable selectivity operates to determine which input lines are to be dealt with at any one time. Strangely, it seems that unconsciousness is the natural state for the CNS to be in. Receptors provide not only specific lines of information but also a constant tonic input which maintains the wakeful state of the CNS. It is thought that a major effect of general anaesthetics is to attenuate sensory input below the point at which the CNS will stay alert, whereupon it lapses into its natural, unconscious state.

1.4.2 Receptor classification

A number of receptor classification systems are available. A semi-functional method was favoured by Sherrington and this is still a reasonable approach today (Table 1.1). Under this system four groups may be identified. Proprioceptors, which signal the position of the body in space or the position of bits of the body relative to other bits, are found in muscles, joints and the vestibular portion of the inner ear. Interoceptors, as their name suggests, signal changes in the internal environment. Examples would include the receptors for blood pressure or pH. Exteroceptors are located in the skin and signal thermal and mechanical changes in the immediate environment.

Table 1.1 — Semi-functional receptor classification. A number of other systems are also in current use

Receptor	Classification
Proprioceptors	Movement and position of body parts
Interoceptors	Internal environment
Exteroceptors	Immediate external environment
Teleceptors	Remote external environment
Nociceptors	Pain

Teleceptors detect changes in the remote environment. These are typified by receptors in the retina of the visual system and those in the inner ear for the auditory system. There may be some justification for adding a fifth group to this basic classification. These are the nociceptors, a specialized subset of receptors for the detection of painful stimuli.

1.4.3 Properties of receptors

It is apparent from the classification that receptors show the property of specificity; that is, there is a type of stimulus to which a particular receptor preferentially responds. The energy form to which a receptor responds best is known as the adequate stimulus. In many cases histological examination of the receptor correlates well with its functional properties, such as the presence of light-sensitive pigments inside photoreceptors. Even within a particular group of receptors there is subspecialization. For instance, in the tongue of the cat thermal receptors respond over distinct temperature ranges of 30 to 32°C or 38 to 40°C. Different groups of photoreceptors show distinct spectral sensitivities.

Under normal operating conditions a particular receptor would be expected to respond to its single adequate stimulus. However, if the input energy is great enough, it is possible to make a receptor respond to other types of stimulation. An example often quoted is that a swift prod in the eye (not an experiment to be recommended) results in the recipient seeing stars, as a result of mechanical stimulation of the photoreceptors.

Before the discovery of the electrochemical nature of neural transmission, it was thought that stimulation of different sensory systems released different types of energy. This was known as the Law of Specific Nerve Energies and was used to explain why stimulation of a particular sensory system always produced the same sensation regardless of what type of stimulus was employed to release that energy. However, since we now know that all our sensory systems utilize the same kind of electrochemical energy, it is thought that it is the central destination and hence the area of brain devoted to processing that information which determines the sensation produced.

That it is the central destination that is of prime importance can be deduced from electrical stimulation of primary cortical receiving areas. For instance, in one or two blind human volunteers arrays of stimulating electrodes have been used to directly stimulate the primary visual cortex in response to a TV camera input. Crude patterns presented to the camera can be seen by the subject.

It is implicit in what has been said that we are capable of detecting many energy types such as heat, light, mechanical energy and sound. Since the nervous system employs only electrochemical energy, the receptor must act as a transducer, changing the energy form to which it is sensitive into electrochemical energy. A variety of mechanisms are utilized to achieve this. Some of them will be described in detail with the appropriate sensory system. One common method is for the impinging energy, perhaps via some intermediate steps, to change the permeability of the receptor membrane to one or more ionic species, which then flow across the receptor membrane down their electrochemical gradients. Since ions carry charge, movement results in a flow of current, the generator current, which establishes a generator potential. The generator potential is a localized potential, restricted to the receptor, and varies in amplitude, between limits, with the intensity of the stimulus producing it. Electrotonic spread of the generator potential then gives rise to action potentials in the afferent axon. These action potentials are of fixed amplitude, their frequency and number encoding generator potential amplitude. The nature of membrane and action potentials are more thoroughly explored in the next chapter.

Receptors show the properties of temporal and spatial summation; that is, generator potentials arising at similar times on the same or different points of the receptor membrane may be addded together in terms of numbers of action potentials produced.

It is often advantageous to be able to spatially localize the stimulus, for instance localize a mechanical stimulus to a particular area of skin. Any one receptor for mechanical energy will only detect mechanical deformation of the skin over a small area. This area over which a receptor responds is known as its receptive field. It is often circular or elliptical in shape for a skin mechanoceptor. The topography of the input from the receptive fields is maintained in the central projection. The concept of receptive field, although easiest to understand in terms of skin mechanoceptors, can be extended to many other receptor types.

If the stimulus to a receptor is maintained, the receptor's response may start to decline: a process known as adaptation. Some adapt very rapidly and are known as phasic receptor types. An example would be rapidly adapting low-threshold mechanoreceptors in the skin. Others adapt very slowly and are known as tonic receptors. An example would be some types of muscle spindle ending.

For many receptor types, considerable modulation of their sensitivity is seen to match their responsiveness to prevailing requirements. This may occur in a variety of ways, for example by innervation of the receptor by efferents from the CNS. In this way receptors in muscle are adjusted to the current operating length of the muscle while those in the auditory system may have their frequency sensitivity adjusted.

2

Neuronal functioning

2.1 AMPLITUDE- AND FREQUENCY-MODULATED TRANSMISSION

The point at which sensory information is received may well be a considerable
distance from that at which it can be processed. Similarly, the command for initiation
of motor or other activity by the CNS may have to travel some distance to the
effector. An enormous amount of information exchange occurs between neurons
within the CNS. Clearly all these instances require the faithful transmission of data
from one point to another. This occurs in the form of electrical impulses, known as
action potentials, which are propagated along the axonal membrane.

2.1.1 Cable properties of the axon

In a good electrical conductor, such as a length of copper cable, a voltage applied at
one end decrements in amplitude very slowly with distance, and thus a reasonably
strong and faithful signal can be transmitted over large distances (Fig. 2.1). The

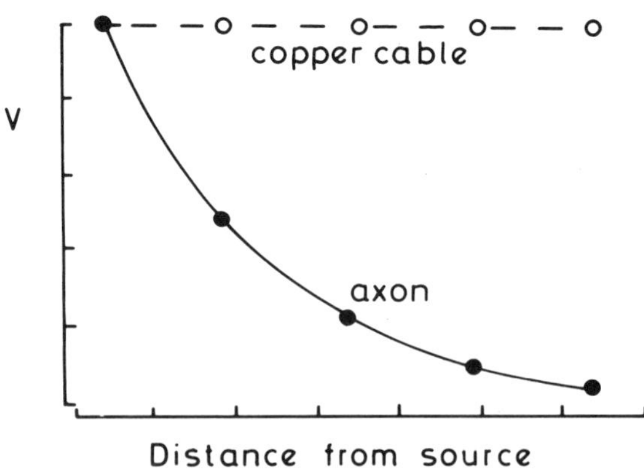

Fig. 2.1 — Showing poor cable properties of axon (dots) relative to copper cable (circles).

axon, however, possesses poor passive cable properties; a voltage applied at one end declines in amplitude very rapidly with distance. Thus a passive transmission system based on changes in signal amplitude (amplitude-modulated) would not be effective for long-distance axonal transmission. Instead, a fixed-amplitude, rapidly attenuated voltage is employed which is constantly regenerated at consecutive close points on the axonal membrane. Information is encoded within the frequency, number and pattern of these action potentials travelling along the axon. This kind of frequency-modulated system transmits information much more faithfully than amplitude modulation and is used extensively in modern computer and communications technology. Let us examine how it is achieved in the axon.

2.2 RESTING MEMBRANE POTENTIAL

Much of the knowledge concerning axonal transmission has come from examination of the axons to the muscle of the mantle of the squid. These are the giants of the axon world being 1 mm in diameter, which greatly facilitates study. By comparison, the largest axons we possess are around 22 μm in diameter. If a pair of recording electrodes are placed across the membrane of the giant axon a steady difference in electrical potential is recorded, the inside being some 70 mV negative with respect to the outside (Fig. 2.2). This is known as the resting membrane potential. Administra-

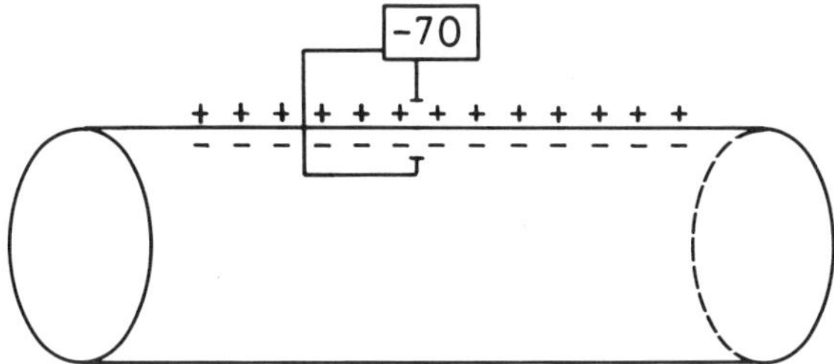

Fig. 2.2 — Resting potential (mV) across the membrane of the giant axon of the squid.

tion of a metabolic poison to the axon results in a slow decline in the transmembrane potential to zero, accompanied by a loss of ability to transmit information. Thus the existence of the resting membrane potential is ultimately dependent upon the expenditure of metabolic energy and itself is a prerequisite for the production of action potentials. Consideration of the resting membrane potential is therefore critical to our understanding of axonal transmission.

2.2.1 Electrochemical gradients

Chemical analysis of the interior of the axon (axoplasm) shows it to have a markedly different ionic composition from the extracellular fluid. Inside the axon is rich in potassium and large organic anions, while the fluid outside the axon contains high concentrations of sodium and chloride ions (Fig. 2.3). It is this differential ionic

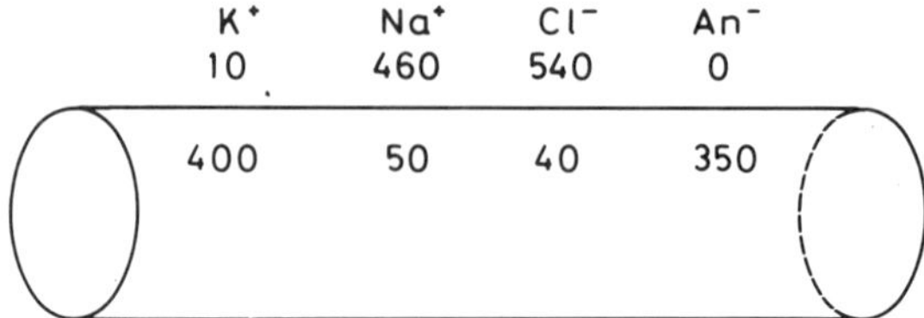

Fig. 2.3 — Differential ionic composition (mM/l) inside and outside the giant axon of the squid. An⁻, organic anions.

distribution which leads to the resting membrane potential. Small particles in solution are in a state of constant diffusive movement. Thus if a concentration gradient exists, there will be a net movement of particles from a region of high to a region of lower concentration, provided there are no permeability barriers obstructing the movement. Such a chemical gradient will apply to ions present in different concentrations on either side of the axonal membrane. Since ions are charged particles, any net movement across the membrane will produce a potential difference across the membrane — a diffusion potential.

2.2.2 Diffusion potential

The axonal membrane is not equally permeable to all ions. The selective permeability of this membrane means that some ionic types, but not others, are allowed to diffuse down their concentration gradients and produce a diffusion potential. Since an ion is a charged particle, its movement will both produce and be influenced by electrical gradients. Thus diffusion of ions across the axonal membrane is the result of a combination of electrochemical forces, together with the properties of the axonal membrane. It is the diffusion of ions across the membrane which produces the resting membrane potential at the equilibrium position between the electrical and chemical gradients.

2.2.3 Nernst equation

It is possible to determine which of the ions is principally or wholly responsible for the resting membrane potential of the axon. A formula can be derived which allows the calculation of a theoretical transmembrane potential from measured chemical gradients. This, in its simplified form, is known as the Nernst equation (Fig. 2.4). If the measured and theoretically derived transmembrane potentials for a particular ion are similar, then the potential could be produced as a result of diffusion of that ion across the membrane down its concentration gradient. The extent to which the maximum theoretical diffusion potential for an ion can develop will be determined by the extent to which the membrane is permeable to that ion.

If we apply this reasoning to the resting membrane potential of the axon, the closest agreement is for potassium ions with a theoretical value of 80 mv negative on the inside with respect to the outside, while, for instance, sodium yields a theoretical value of +55 mv inside relative to outside. Thus it appears, and has been shown, that the major contribution to the resting membrane potential is diffusion of potassium

NERNST EQUATION

$$E_K = \frac{RT}{zF} \ln \frac{[K]_o}{[K]_i}$$

or extended to
GOLDMAN EQUATION

$$E = \frac{RT}{zF} \ln \frac{P_K[K]_o + P_{Na}[Na]_o + \ldots}{P_K[K]_i + P_{Na}[Na]_i + \ldots}$$

WHERE

E, membrane potential
E_K, equilibrium potential for potassium ions
R, gas constant
T, absolute temperature
z, ionic charge
F, Faraday's Constant
$[K]_o$ and $[K]_i$, ionic concentration of potassium outside and inside axon
$[Na]_o$ and $[Na]_i$, ionic concentration of sodium outside and inside axon
P_K and P_{Na}, relative permeabilities to potassium and sodium (in resting state 1 and 0.013 respectively).
ln, natural logarithm

Fig. 2.4 — The Nernst and Goldman equations.

ions down their chemical concentration gradient from the inside to the outside. By contrast, the membrane permeability to sodium ions in the resting state is very small. The Nernst equation can be extended to take account of the small contributions of other ions, such as sodium and chloride, to the resting membrane potential. It is then known as the Goldman equation and comes even closer to the actual measured membrane potential.

2.2.4 Ionic channels in the membrane — ionophores

Chemically, the axonal membrane is thought to consist of a phospholipid bilayer together with protein, some absorbed onto the surface, some embedded within the bilayer (Fig. 2.5). Experiments with artificial membranes show that ions pass

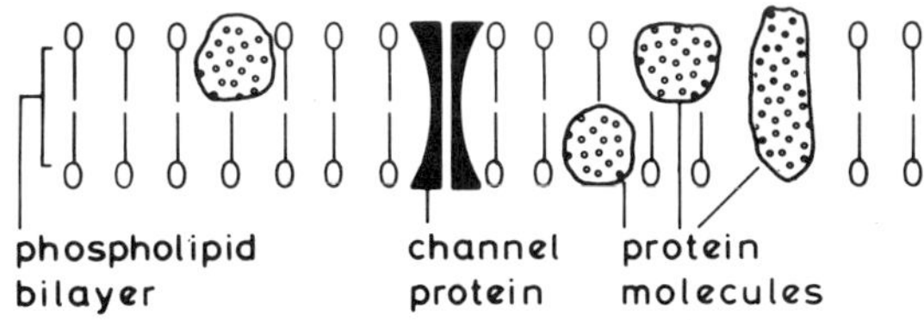

Fig. 2.5 — Structure of cell membrane of neuron. Such diagrams tend to suggest a rigid, static barrier layer. In reality, however, the membrane is a highly fluid, dynamic structure with the various protein components floating around in the liquid crystalline phospholipid.

through the lipid phase only with great difficulty. However, if a second type of molecule, such as a cyclic peptide antibiotic like valinomycin, is incorporated into the artificial membrane, ionic permeability is greatly increased. It is thought that these molecules form suitable aqueous channels for ions through the membrane. Some common antibiotics may act in this way by destroying the ionic balance of the micro-organisms they attack. These ionic channels are often referred to as ionophores.

In a similar fashion, some of the protein of the axonal membrane is in the configuration of helical molecules forming aqueous channels for potassium ions to pass through, and one type is permanently open. We shall meet a second variety, the gated channel, shortly.

2.2.5 Channel specificity

Clearly, these channels are ion-specific, since sodium ions cannot pass through the potassium channel. One factor in channel specificity is ionic size. In solution, ions are hydrated; that is, they drag around with them a large shell of water. The sodium ion hydration shell is larger than that of the potassium ion and may be too large to pass through the potassium channel.

2.3 ACTION POTENTIAL GENERATION

Development of the action potential at a point on the membrane arises from a reversal of the membrane potential so that it momentarily becomes positive on the inside with respect to the outside by some 40 mv (Fig. 2.6). It will be recalled that this

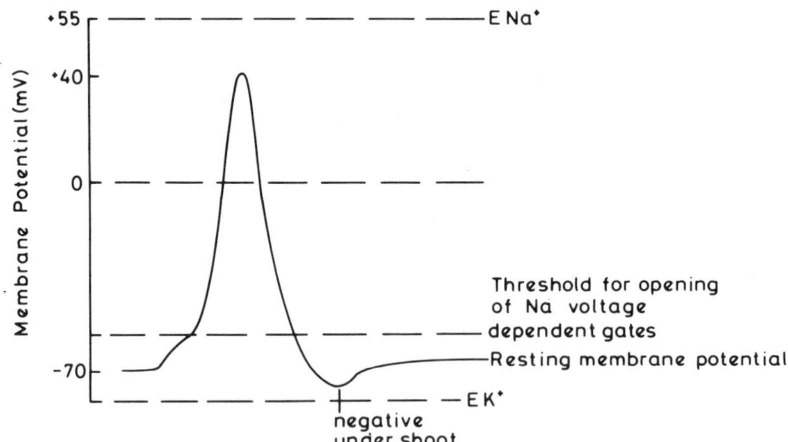

Fig. 2.6 — The development of the action potential across the membrane. Note the upswing heads towards the sodium equilibrium potential and the downswing back towards the potassium potential.

value is close to the membrane potential which would occur as a result of a diffusion potential for sodium ions. Thus it appears that the action potential results from a sudden switching on of sodium ion permeability.

2.3.1 Voltage-dependent channels

One suggestion for the nature of the membrane channels for sodium is that they may be molecules carrying a dipole, one end having a slight negativity, the other a slight positivity (Fig. 2.7). These are voltage-dependent channels; whether they are open

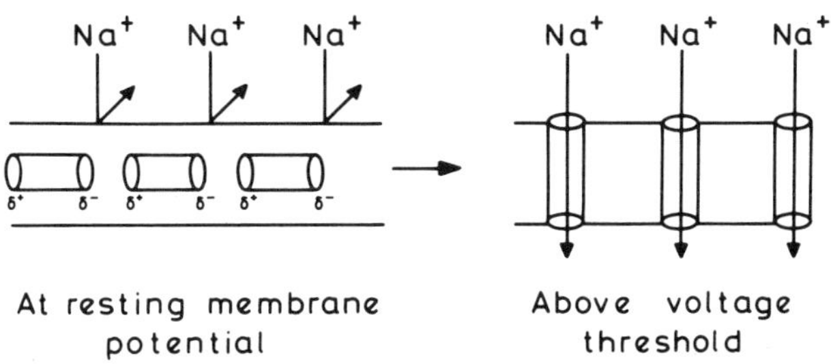

Fig. 2.7 — Orientation of dipole-carrying voltage-dependent sodium channels at different membrane potentials.

or not depends upon the membrane potential. At the resting membrane potential they are thought to line up head-to-tail along the length of the membrane and are therefore in the closed position. A small depolarization (by about 15 mv closer to zero), however, results in them flipping through 90° so that they now span the membrane and are open to sodium ions which can then diffuse across. Following this brief opening the voltage-dependent gates again close and the potential heads back towards the equilibrium potential for potassium ions, in some nerves briefly overshooting the resting membrane potential due to the transient opening of gated potassium channels in addition to the continuously open potassium channels.

2.4 ACTION POTENTIAL PROPAGATION

The reversal of potential at one point on the membrane causes local currents to flow which depolarize an adjacent point sufficiently to open the voltage-dependent sodium gates. An action potential is thus generated at that point and so on. The regeneration of the action potential at consecutive points along the membrane results in its propagation along the axon. For a short period after production of an action potential the sodium gates cannot be reopened. During this time, the absolute refractory period, the nerve cannot produce another action potential at this point on the membrane. This causes unidirectional propagation along the axon and imposes an upper limit on the action potential frequency.

2.4.1 Saltatory conduction

The regeneration of action potentials at a large number of consecutive points on the axonal membrane has two major disadvantages. First, since each action potential requires a finite time to develop, propagation along the nerve is quite slow, which of course means that the conduction velocity of the nerve is low. Second, since each action potential results in the intra-axonal gain of a small amount of sodium and the loss of a small amount of potassium, considerable metabolic energy must be expended to maintain the concentration gradients. Nerve conduction under these conditions is both slow and metabolically wasteful. One evolutionary solution to improving conduction velocity is to increase the axonal diameter, which decreases the longitudinal resistance improving propagation rate. This ploy has been carried to its extreme in the giant axon of the squid, where rapid activation of the propulsive mantle will endow a survival advantage.

Another significant development is the wrapping of the axon in a sheath of the fatty substance myelin. These myelin segments are around 1 mm in length, each segment being secreted by a glial cell. The axonal membrane is left exposed between adjacent segments (nodes of Ranvier) which are the only points at which the action potential can be regenerated. Poor though the cable properties of the axon are, there is sufficient signal strength left from an action potential at one node to provide local currents to depolarize the next node to the point at which the voltage-dependent gates open to regenerate the action potential, which depolarizes the next node and so on. This type of propagation, in which the action potential appears to jump from node to node, is known as saltatory conduction (Fig. 2.8). The use of myelin to

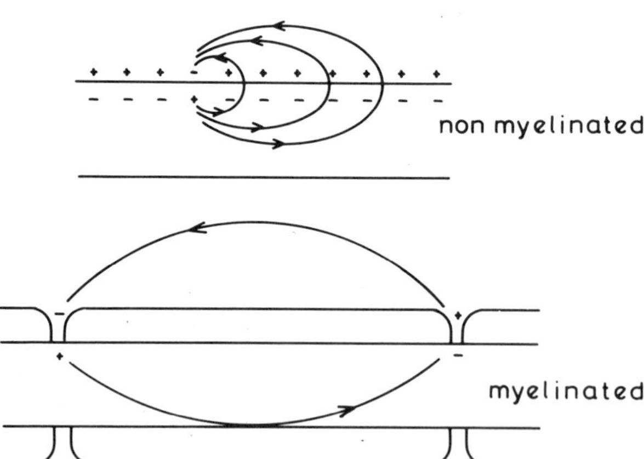

Fig. 2.8 — Local current flow in myelinated and non-myelinated axons. These local currents depolarize the next available part of the axonal membrane sufficiently to open the voltage-dependent sodium gates.

restrict the number of regeneration points and reduce current leakage out of the axon between nodes greatly increases the propagation velocity. Further, myelinated fibres do not utilize gated potassium channels during repolarization. This, together with restriction of the number of regeneration sites, means less pumping of ions is required to maintain concentration gradients and therefore greater metabolic efficiency. The human nervous system contains a whole spectrum of nerve fibres from small unmyelinated to large myelinated. They may usefully be classified according to diameter and conduction velocity (Table 2.1). The systems using the various fibre types will be described in other chapters.

Table 2.1 — Nerve fibre classifications based on diameter and conduction velocity. The upper one (types A–C) is the more widely used, while the lower one (types I to IV) was evolved by those groups studying muscle afferents

Fibre Description	Diameter (μm)	Conduction Velocity (ms^{-1})	Myelination	Typical representative
Nerve fibre classification				
Aα	12–22	70–120	Yes	Cutaneous tactile afferent
Aβ	5–12	40–70	Yes	Cutaneous tactile afferent
Aγ	3–6	15–40	Yes	Muscle spindle efferent
Aδ	1–5	12–30	Yes	Fast pain afferent
B	<3	2–15	Yes	Preganglionic sympathetic
C	0.4–1	0.6–1.6	No	Slow pain afferent
Alternative system for muscle afferents				
Ia	12–22	72–120	Yes	Primary annulospiral ending on muscle spindle
Ib	Slightly smaller but extensive over-lap with Ia		Yes	Golgi tendon organ
II	6–12	36–72	Yes	Secondary ending on muscle spindle
III	1–6	6–36	Yes	Joint receptors
IV	0.4–1	0.6–1.5	No	Deep pain

2.5 SODIUM–POTASSIUM PUMP

The question of how the chemical gradients are established in the first place still needs to be addressed. Production of the resting potential results in constant leakage of potassium from the axon, while a small amount of sodium leaks in. During each action potential a small amount of sodium is gained and in some axons there is further loss of potassium. In the axonal membrane is a linked sodium–potassium pump, which, utilizing metabolic energy derived from adenosine triphosphate (ATP), actively pumps back into the axon the potassium which diffuses out in exchange for sodium ions which are pumped out (Fig.2.9). The pump is running all the time to

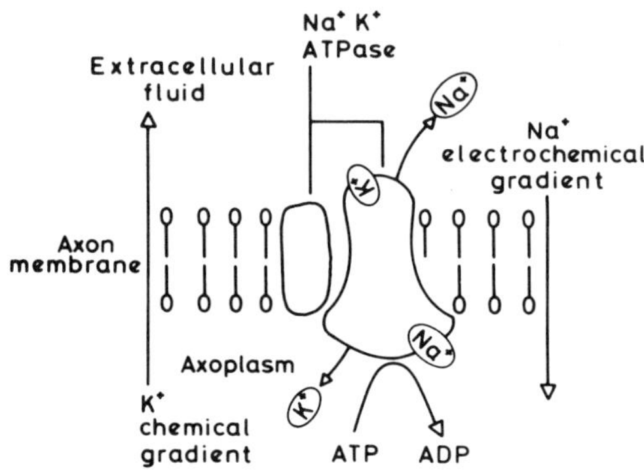

Fig. 2.9 — Linked, energy-consuming membrane transport system for sodium and potassium ions. The pump may work at different rates and exchange varying ratios of potassium and sodium ions according to circumstances. Up to three sodium ions may be pumped out for every two potassium ions brought in.

maintain the appropriate resting ionic gradients. The rate at which it operates is geared to respond to prevailing conditions ranging from a basal rate when no action potentials are being generated to a much faster rate when neural traffic is heavy.

2.6   SYNAPTIC TRANSMISSION

The neuroanatomist Ramon y Cajal first observed that physical continuity is broken at many points in the nervous system, leaving a gap known as a synaptic cleft of between 10 and 40 nm between one neuron and the next (up to 1000 nm at the neuromuscular junction). Any one neuron may receive synaptic input to its dendrites, soma and terminal parts of its axon, and may give synaptic input to many other neurons. The opportunities for exchange and integration of information therefore are staggering. Synapses form a link from one neuronal element to the next and should not be thought of as static devices, but rather as dynamic elements which are being continuously modified, with important consequences for the information they transmit. Modern theories of memory involve consideration of synaptic changes, although this aspect will be developed in Chapter 12.

2.6.1   Nature of transmission — chemical conduction

At most synapses, arrival of the action potential at the axon terminal leads to a release of a chemical messenger from the synaptic terminal. This diffuses across the synaptic gap — a fairly rapid process over such tiny distances — and produces an effect on the postsynaptic membrane (Fig. 2.10). One common effect is for the

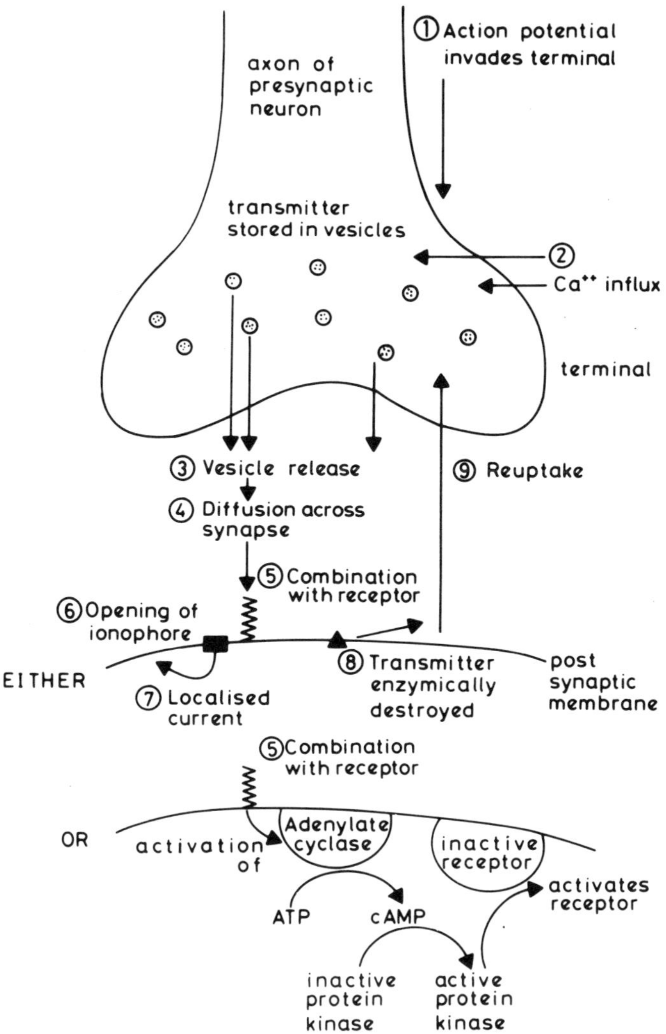

Fig. 2.10 — Major stages in synaptic transmission. Postsynaptic steps most often lead directly to membrane permeability changes. Less frequently a secondary messenger may be activated. Stage 8 (enzymatic) degradation, may not occur, rather the whole transmitter molecule may be taken back up into the presynaptic terminal.

postsynaptic membrane to become depolarized. A great many different chemical messengers are thought to be used by the nervous system. The present count is over 50 and rising.

Many neurons produce one or more branches off the major axon. For example the motoneurons in the spinal cord not only supply skeletal muscle peripherally but also give off short branches centrally which innervate inhibitory feedback interneurons. Sir Henry Dale suggested that it was likely that all terminals of a particular neuron released the same chemical messenger. This useful working hypothesis was ossified by others into the Dale Principle — that a single neuron released only a single transmitter. However, more and more instances are coming to light of neurons which release two or more chemical messengers. Let us consider in more detail the process of synaptic transmission. The transmitter substance is synthesized in the cell soma and transported within the microtubule system of the axon. On reaching the axon terminal it is stored in small amounts inside membranous vesicles around 25 nm in diameter. Not all vesicles contain the same amount of transmitter substance. Transmitter release involves the fusion between the vesicle and axonal terminal membrane followed by rapid breakdown of the common membrane, resulting in the transmitter being released into the synaptic cleft.

2.6.2   Excitation–secretion coupling

The link between terminal excitation by the action potential and transmitter release is thought to involve calcium ions. In addition to the kind of ionic channels already described, the axon terminal contains gated channels for calcium ions. These are voltage-dependent and open on arrival of the action potential. Since the extracellular concentration of calcium ions is much higher than the intracellular concentration, calcium ions flood into the nerve terminal. The presence of free ionized calcium in the terminal causes vesicles to move to and fuse with specialized regions of the axonal membrane, possibly by activation of contractile protein molecules similar to those in muscle. Transmitter is then released into the synaptic cleft. The number of vesicles released is proportional to the amount of free ionized calcium present in the terminal, which in turn depends on the number, frequency and (as we shall see in section 2.6.8) amplitude of the action potentials arriving. The calcium flooding in as a result of any particular action potential is rapidly removed by being taken up by specialized proteins and by cell organelles, and by being pumped out.

2.6.3   Postsynaptic events — permeability changes and second messengers

On arrival at the postsynaptic membrane the transmitter molecule combines with a specific receptor molecule. The consequences of this interaction vary depending on the transmitter involved and the site of action. The most usual effect is for the receptor to control an ionic channel (ionophore). Formation of the transmitter/ receptor complex alters the configuration of the ionophore, changing the ionic permeability of the membrane in that region and resulting in localized current flow. A less common mechanism is for the transmitter/receptor complex to initiate a complex series of biochemical reactions involving second messengers which may alter the metabolism of the innervated cell or change the sensitivity of some subset of receptors. Some of these permeability or other changes are described in detail below.

2.6.4   Excitatory postsynaptic potential (EPSP)

The EPSP is brought about by increased membrane permeability to sodium and potassium ions leading to a non-propagated localized depolarization of the mem- brane (Fig. 2.11). Typically several thousand such synaptic contacts are available on the neuronal membrane. They are integrated by the neuronal membrane and, if the resultant excitation is sufficient, cause the generation of axon potentials from the initial segment. The number and frequency of action potentials produced is pro-

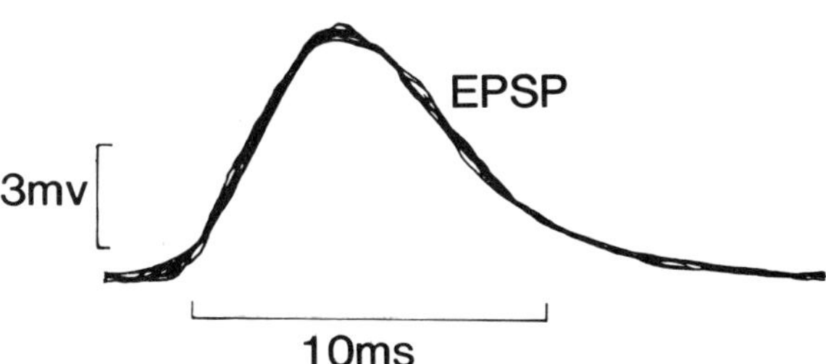

Fig. 2.11 — EPSP recorded from a motoneuron.

portional to the net integrated excitation. For example, stimulation of muscle spindle Ia afferents produces EPSPs in the motoneurons of their muscle of origin. A sufficiently large, synchronized volley will cause motoneuron discharge and reflex contraction of the muscle. This is the basis of the muscle stretch reflex.

2.6.5 Inhibitory postsynaptic potential (IPSP)

The IPSP is brought about by increased membrane permeability to chloride ions, leading to a net influx of this ionic species, hyperpolarizing the membrane at this point, carrying it further from its threshold, and making it less likely to fire (Fig. 2.12). A good example is furnished again by the spinal motoneuron. The efferent

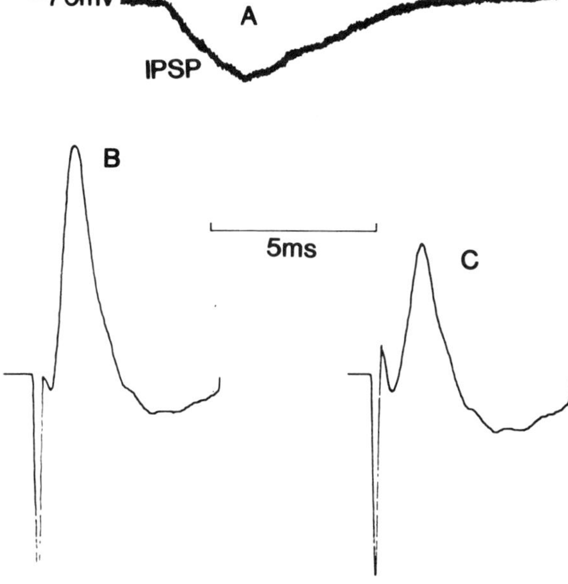

Fig. 2.12 — The upper part of the figure (A) shows an IPSP recorded from a motoneuron. In the lower part of the figure excitability of motoneuron pool tested by antidromic stimulation is shown under control conditions (B) and while motoneuron membranes are subjected to IPSPs (C). In the presence of the inhibition fewer of the motoneurons discharge to the test stimulus and a smaller volley is evoked.

carrying the excitatory impulses to muscle gives off a collateral within the spinal cord which innervates a small interneuron known as a Renshaw cell. Activation of these cells results in an IPSP in the motoneuron and surrounding motoneurons. Under normal operating conditions it is thought to exert a form of surround inhibition to limit the rate of firing of any particular motoneuron and sharpen up motor output, in addition to other important effects.

2.6.6   End-plate potential (EPP)

A motor axon entering a muscle splits into a number of terminals which innervate several separate muscle fibres. A motoneuron together with the muscle fibres it innervates is known as a motor unit. There is enormous variation between muscles in the number of muscle fibres controlled by a single motoneuron and this is inversely related to the fineness of control we can exert over that muscle. This ranges from five muscle fibres per motoneuron in the extraocular muscles to 1000 per motoneuron in the temporal muscle.

There is a gap of around 1000 nm between the motor nerve ending and the skeletal muscle fibre membrane. The transmitter substance here is acetylcholine. Curare, a potent blocker of acetylcholine action at the skeletal neuromuscular junction, is the active principle of the poison-tipped arrows used by South American Indians. Its use results in muscular paralysis of the prey. Curare-like substances are also useful clinically to eliminate muscular reflexes during surgery. Since many anaesthetics are toxic, simultaneous use of muscle blockers allows a much lower dose of anaesthetic to be used. However, anaesthetic level must be carefully monitored, since a paralysed artificially respired patient is incapable of signalling their level of awareness.

After diffusing across the synaptic cleft the acetylcholine reacts with receptor proteins present on the specialized end-plate region of the muscle membrane and opens large-diameter non-specific channels here. The various ions present (sodium, potassium and chloride) diffuse down their electrochemical gradients, causing a localized depolarization of the membrane — the end-plate potential (Fig. 2.13). This

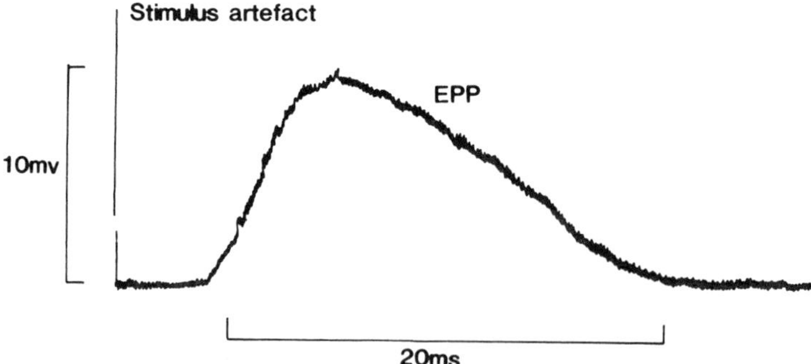

Fig. 2.13 — EPP recorded from the end-plate region of a frog sartorius muscle to stimulation of its motor nerve. The evoked EPP would normally trigger action potentials in the adjacent muscle membrane. The much larger action potentials would then obscure the EPP. To prevent this the preparation is curarized to reduce the EPP amplitude below the level at which an action potential is produced.

initiates propagating action potentials along the muscle membrane. The acetylcholine is then destroyed by a neighbouring enzyme complex containing the enzyme cholineesterase.

2.6.7 Second messengers
The mechanisms so far described result from changes in membrane permeability. In some cases the combination of the transmitter with the receptor does not open ionic channels in the membrane, but, via coupling proteins, switches on an intracellular biochemical sequence. This sequence may result in the activation of some other membrane receptor or switch the metabolic direction of the neuron.

2.6.8 Presynaptic inhibition and facilitation
Inhibition produced by hyperpolarization of the postsynaptic element has already been described. This carries the membrane further from its threshold, making it less likely to fire. Another type of inhibition has been identified with a much longer time course and a site of action on the presynaptic terminal. It is known as presynaptic inhibition. The relationship between intraterminal free calcium concentration and neurotransmitter release is very steep, with small reductions in calcium influx resulting in large decreases in neurotransmitter released. Recordings from nerve terminals which are being inhibited in this way show that there is a decrease in the action potential amplitude, which in turn reduces the calcium influx and consequently the amount of neurotransmitter released (Fig. 2.14).

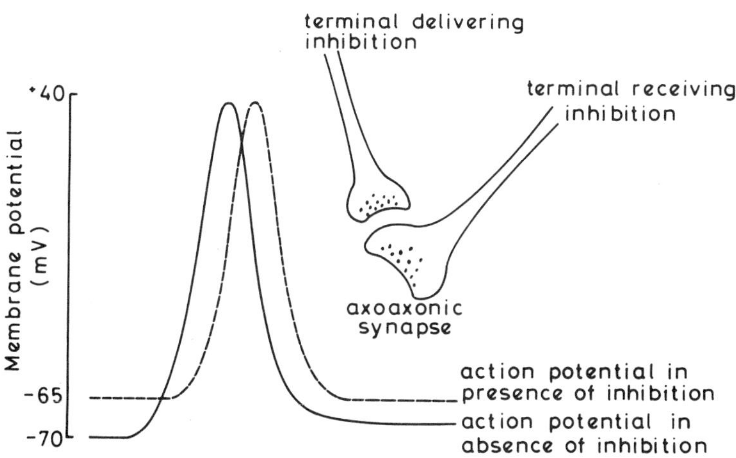

Fig. 2.14 — Effect of presynaptic inhibition on the amplitude of the action potential at the axonal terminal.

The two points at which the action potential could be varied would be either by moving the starting point nearer to, or the peak further from, zero. Since the peak represents a rapid swing towards the sodium equilibrium potential, any changes here would seem unlikely. The closer a nerve membrane is to its threshold, the less electrical current need be supplied to bring it to its firing threshold. Wall utilized this phenomenon in showing that spinal dorsal root nerve terminals were easier to fire by electrical stimulation while they were simultaneously subjected to presynaptic inhibition by stimulation of neighbouring fibres. Thus presynaptic inhibition operates by making the action potential start from a point closer to zero. Presynaptic inhibition is used in many parts of the CNS. When, as here, it is elicited at the terminals of the sensory afferent fibres, it is often known as primary afferent depolarization (PAD).

Subsequent work seemed to suggest that presynaptic inhibition was a rather non-specific phenomenon. Thus stimulation of any part of a dorsal root would induce PAD in all the surrounding inactive terminals. For this reason it was proposed that it results from potassium leakage from active fibres depolarizing the surrounding terminals within diffusive range. However, measurement of the amount of potassium actually released under these circumstances suggests that it plays only a minor role.

More recent evidence has uncovered the production of more precise patterns of inhibition. Physical evidence has been adduced in the form of axon terminals forming synapses on other terminals — axo-axonic synapses. The transmitter γ-aminobutyric acid (GABA), in addition to its more familiar action of inducing IPSPs, has also been shown to produce PAD. Thus it appears that chemically mediated synapses are the major factor in presynaptic inhibition.

Axo-axonic synapses can also be used to produce presynaptic facilitation. Instead of depolarizing the terminal on which they are acting, they hyperpolarize it, that is move the starting point for the action potential further from zero. Any action potential now invading the terminal is increased in amplitude and more transmitter is released. The ionic basis of primary afferent hyperpolarization is unclear.

2.6.9   Electrical synapses

Although most neuronal connections employ some kind of chemical transmission, a small number utilize electrical conduction. In order for electrical conduction to be possible between one neuronal element and the next, sufficient current has to cross the synapse to effectively depolarize the postsynaptic membrane. There are considerable problems associated with such transmission at all but a small number of synapses. For example the presynaptic element supplying the current is often small and can therefore supply only small amounts of current. Current will tend to leak away during its passage across the synapse. The postsynaptic element may be much larger and may have a high electrical resistance. However, even with all these problems electrical transmission does occur at some sites, for instance through special tight junctions between cardiac muscle cells. Electrical transmission has also been identified in some parts of the adult mammalian CNS and is probably more prevalent at the immature synapses of neonate and young animals.

3

Visual system

The human eye is a collection of some 126 million receptors specialized for the processing of part of the electromagnetic spectrum between wavelengths 450 and 700 nm (Fig. 3.1). These receptors are aggregated in the retina, which embryologi-

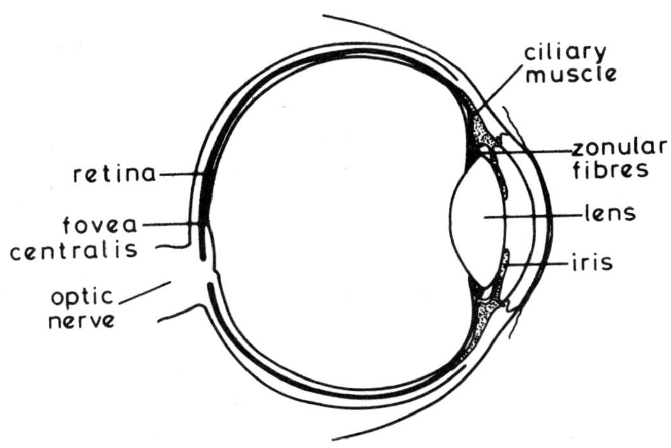

Fig. 3.1 — Simplified diagram of the eye. Note insertion of zonular fibres around the edge of the lens.

cally is developed from the forebrain. Strangely, the receptors are found in the deepest part of the retina. To reach them, therefore, light must traverse the rest of the retina. In front of the retina lie systems to focus and adjust the amount of light entering the eye.

3.1 IMAGE FORMATION BY THE EYE — TRIPLE RESPONSE

Certain mechanical changes must occur in the eye to allow retinal focussing of the image under different conditions. Three major changes occur when we focus on an object brought close to the near point. They are accommodation, miosis and ocular convergence, collectively known as the triple response.

3.1.1 Accommodation

When we look at a distant object, some 36 of the 42 dioptres (D) of focussing power required is provided by the cornea. However, variations in focussing power for objects at different distances is provided by the lens. While its contribution for objects at a distance is small, it plays a much bigger part when objects are close to the eye. For example, when an object is at the near point, the minimum distance from the eye at which focussing can be achieved (about 8 cm in the young, healthy eye), it provides up to 24 of the 60 D required. This adjustment of the lens to focus at the near point is known as accommodation, and is achieved by the ciliary muscle, the zonular fibres of which are inserted around the lens circumference holding the elastic lens under tension. Ciliary muscle contraction relieves this tension, allowing the lens to assume a more spherical shape, increasing its focussing power.

3.1.2 Miosis

The amount of light reaching the receptors is determined by the size of the pupil, the diameter of which can be varied between 2 and 8 mm by the action of the muscle fibres in the surrounding iris. Pupillary constriction (miosis), occurs when focussing on a near object. In addition to reducing the total light input, thereby reducing the possibility of saturating the receptor mechanism, it also restricts the extent of retina illuminated to a specialized region, the fovea centralis, and, by allowing light to pass through only the central portion of the lens, reduces problems associated with spherical and chromatic aberration. A widely dilated pupil, on the other hand, has long been associated with sexual attractiveness, and historically was achieved by blocking the sympathetic cholinergic input to the circular muscle of the iris using atropine (belladonna). Unfortunately this would seriously impair vision, which would presumably have led to some interesting liaisons.

3.1.3 Convergence

Control of the amount of medial rotation of the eyes by means of the extraocular muscles allows us to maintain a binocular view of objects, which is important in depth perception.

3.2 RECEPTOR ACTIVATION

Light is thus brought to focus on the receptor layer of the retina. Two types of photoreceptor have been identified. They are known as rods and cones and fill different, though complementary, functions in vision. Each receptor of either sort is divisible into inner and outer segments, a nuclear region and a synaptic element. The outer segment is embedded in the melanin-containing pigment layer of the retina. This darkly pigmented layer prevents reflection of the light back through the retina and the cells here perform a number of metabolic services for the photoreceptors. The photoreceptor contains light-sensitive pigment stacked in disc-like structures in the outer segment.

3.2.1 Dark current

The neurophysiology of the retina shows some fascinating and unique properties, one of which is the amount of processing which occurs by the electrotonic spread of slow waves rather than by propagation of action potentials. Another is the transduction process itself. If a microelectrode is advanced through the receptor layer while keeping the retina in total darkness, something rather interesting is observed (Fig. 3.2). There is a constant flow of current out of the inner segment into the outer

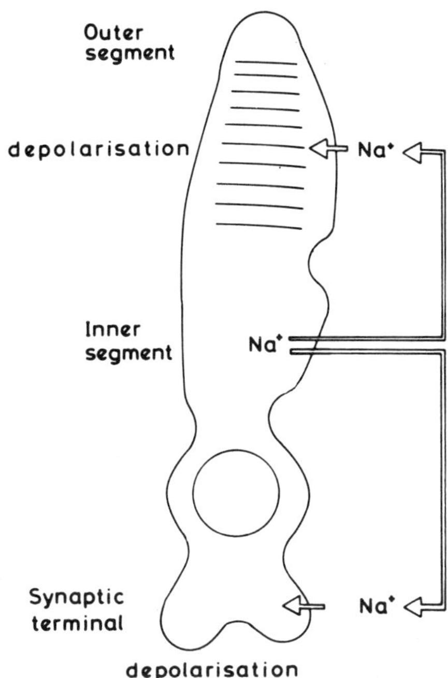

Fig. 3.2 — Flow of dark current, carried by sodium ions, from the inner segment to the outer segment and synaptic terminal region.

segment and synaptic region. This 'dark current' is carried by Na^+ ions, and depolarizes the membrane in the regions of the outer segment and synapse. Thus, whereas receptors of many other modalities depolarize upon stimulation, the photoreceptors are depolarized in the absence of stimulation. Presentation of a flash of light results in a rapid cessation of the dark current and therefore of the depolarization of the receptor membrane. This results in the receptor becoming hyperpolarized with respect to its unstimulated state. One can only speculate why photoreceptors work in this way. It may be that a much quicker response can be registered by turning the conductance mechanisms off rather than switching them on, which would be of considerable benefit when one considers the rapid and transient nature of visual stimuli. Careful inspection of this hyperpolarizing receptor potential gives us an idea of a possible link between intracellular events and subsequent transmembrane consequences.

3.2.2   Receptor potential

The receptor potential is divisible into early (ERP) and late (LRP) components (Fig. 3.3). The ERP is almost instantaneous, and rather insensitive to anoxia or anaesthe-

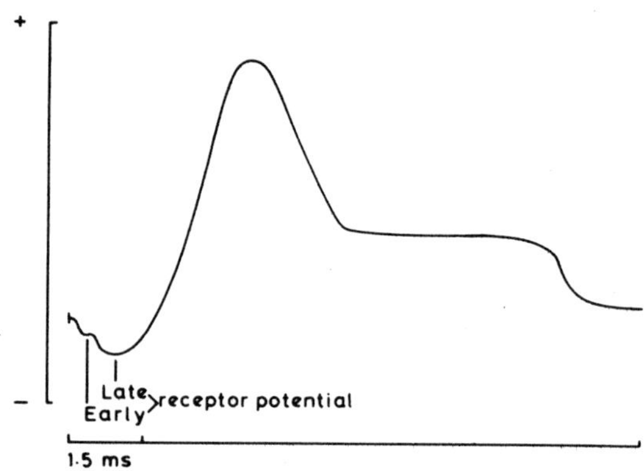

Fig. 3.3 — Electroretinogram recorded upon illumination of retina. The sequence of waveforms represents responses from different neuronal types and support cells. The earliest response seen is the negative going receptor potential.

tics. It is thought to result from conformational changes in the photopigment triggered by absorption of one or more photons of light. The LRP has a latency of around 1.5 ms, is anoxia- and anaesthetic-sensitive and is considered to reflect the turning off of the membrane conductance to Na^+ ions. The link between the ERP and LRP is via a series of biochemical intermediaries (Fig. 3.4). Formerly this was thought to involve the release of bound Ca^{2+} ions from the light-activated photopigment which then blocked the membrane Na^+ gates. Recent evidence suggests that the Na^+ gates are maintained open by intracellular cyclic guanine monophosphate (GMP). Photopigment activation causes a lowering of the concentration of cyclic GMP and therefore closing of the membrane Na^+ gates.

3.3   INTRARETINAL NEURONAL INTERACTIONS

The photoreceptors are in synaptic contact with bipolar cells which in turn synapse with the ganglion cells (Fig. 3.5). The ganglion cells are the output neurons of the

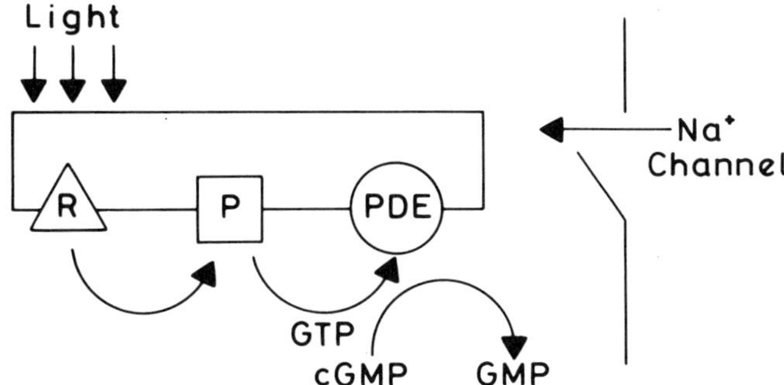

Fig. 3.4 — Sequence of biochemical changes in receptor to a flash of light. R, rhodopsin; P, GTP binding protein, PDE, phosphodiesterase enzyme complex; GTP, guanosine triphosphate; cGMP, cyclic gaunosine monophosphate; Na$^+$ channel, cGMP-controlled Na$^+$ gates in receptor membrane. (After Attwell, D. (1985) *Nature* **317** 14–15).

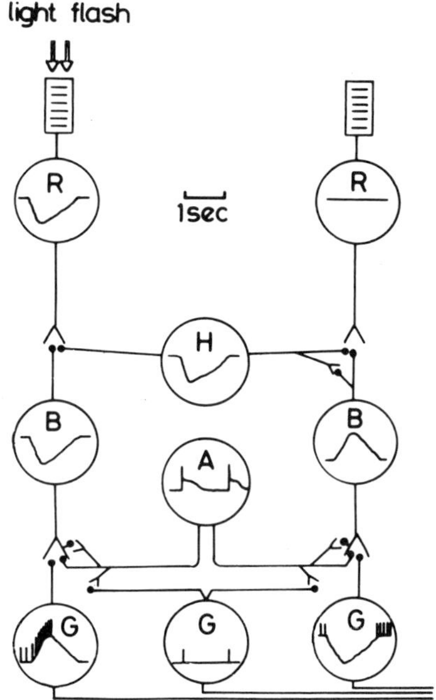

Fig. 3.5 — Retinal neuronal responses when receptor (R) on left illuminated by brief flash of light. H, horizontal cell; B, bipolar cell; A, amacrine cell; G, ganglion cell. Depolarizations are represented by upward deflections, hyperpolarizations by downward deflections. (After Dowling, J. E. and Werblin, F. S. (1969) *J. Neurophysiol.* **32** 315–338).

retina.Their axons are the optic nerve fibres. Within the retina a number of other neuronal types mediate a variety of surround and feedback effects. These include horizontal, interplexiform and amacrine cells.

The amplitude of the receptor potential is proportional to the stimulus intensity. It is a slow electrotonically transmitted wave which can spread to adjacent receptors. No action potentials are produced by the receptor. The hyperpolarization of the receptor causes a similar hyperpolarization of the bipolar and horizontal cells with which it synapses. Here too no action potentials can be recorded. The hyperpolarization of the bipolar cell, however, produces an increase in the frequency of action potentials of the ganglion cells and these action potentials are then transmitted centrally in the optic nerve fibres.

A number of lateral and feedback mechanisms are in evidence in the retina important for sensitivity adjustment and contrast enhancement. For instance the receptor potential, originating in a particular photoreceptor, can spread to neighbouring receptors. At the level of the outer plexiform layer, where the receptors synapse with the bipolar cells, are found the horizontal cells, which also receive synaptic input from the receptors, and give synaptic input to neighbouring receptors and to the bipolar cells of neighbouring receptors. This complex, three-element synapse between receptor, bipolar and horizontal cell is often known as a Triad. Activation of a particular receptor causes hyperpolarization of its horizontal cells, which induces a depolarization in neighbouring bipolars, leading to a decrease in discharge frequency of their ganglion cells . This is a variety of surround inhibition, which is encountered in many guises throughout the nervous system. Here, presumably, it is used to improve and adjust spatial contrast. Feedback occurs between the inner and outer plexiform layers by means of the interplexiform cell, which may be important in adjusting sensitivity of the system to different lighting conditions.

3.4 DUPLICITY THEORY OF VISION

It is thought that the rod and cone receptors fill different, though complementary, functions in vision and this forms the basis of the duplicity theory.

In each retina there are many more rods (120 million) than cones (6 million) and they differ in spectral sensitivity, retinal distribution and connectivities with interneurons. Each rod contains the same photopigment, rhodopsin, which is maximally sensitive in the blue/green part of the spectrum. Each cone, however, contains one of three photopigments maximally sensitive in the red (570 nm — erythrolabe), green (535 nm — chlorolabe) or blue (445 nm — cyanolabe) part of the spectrum. The photopigments are complex molecules consisting of a light-sensitive substance known as retinene, structurally related to vitamin A, together with a protein molecule. The light-sensitive portion is the same in each case. Differences in the protein component confer differential spectral sensitivity.

While rods predominate in the peripheral parts of the retina, the fovea receptor population consists entirely of small, tightly packed cones approximately 1.5 μm in diameter. It is this portion of the retina which is used for detailed visual examination of objects.

Since there are 126 million receptors in the retina but only 1 million fibres in the optic nerve, there is clearly a considerable amount of convergence onto the output cells. Here again there are dramatic differences between rods and cones. Whereas in the rod system many rods converge on a single bipolar cell and a number of bipolars converge on a single ganglion cell, much less convergence occurs in the cone system. This means that the receptive field of a rod-serving ganglion cell is much larger than that of a cone-serving one. Since stimulation of separate points within the receptive field of the same ganglion cell will not be visually differentiable as separate points, the ability to see detail, i.e. visual acuity, is inversely related to receptive field size and is therefore much greater for the cone system. The considerable convergence in the rod system, although resulting in a sacrifice of acuity, allows the spatial summation of light input from many rods. This feature, together with a 10 times greater light sensitivity of the rod over the cone confers the ability to see at much lower light levels than is possible in the cone system. Thus rods give us monochromatic, low-acuity vision in poor light, whereas cones confer on us coloured, high-acuity vision in good light.

Phylogenetic trends may be observed in relative proportions of rod and cone receptors, nocturnal animals tending to possess an even greater preponderance of rod receptors.

3.5 COLOUR VISION

In the nineteenth century a number of theories of colour vision were formulated. Thomas Young in 1801 first showed that any colour could be produced by various mixes of light of the three primary colours, red, green and blue. Helmholtz later applied this train of thought to interpretation of colour vision and postulated that three sorts of colour receptors with the appropriate spectral sensitivities would be necessary. This is the Young–Helmholtz trichromatic theory of colour vision. That the three basic colour receptor types have been shown to exist was a triumph for theoretical science (Fig. 3.6). Hering's opponent process theory took a rather

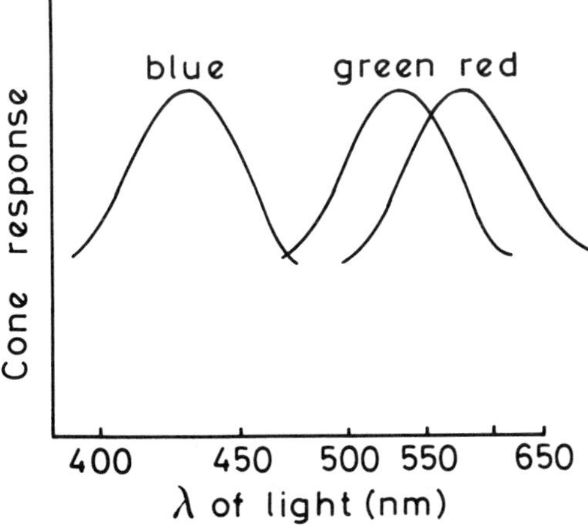

Fig. 3.6 — Spectral sensitivities of cone photoreceptors.

different line in proposing that receptors for pairs of colours worked in an antagonistic manner. As we shall see later, both these theories contain elements of truth. There are indeed three basic colour receptor types while, mutual inhibition between certain colour pairs can be identified during retinal colour processing.

3.6   RETINAL OUTPUT — GANGLION CELLS

The point of final output from the retina is the ganglion cell, whose axons constitute the optic nerve fibres. Microelectrode recording of changes in action-potential frequency produced by these neurons in response to shining small spots of light on the retina has been a valuable tool for the study of retinal mechanisms.

3.6.1   Properties of ganglion cells

In the absence of stimulation, ganglion cells show a low rate of spontaneous discharge. They possess an approximately circular receptive field; that is, they respond to stimulation of an approximately circular area of retina. By the nature of their response they may be divided into W, Y and X types. The W type seem to be signalling light intensity, responding to light or dark spots projected onto their receptive fields, and may be part of the luminosity detection system providing the information for regulating light input by adjusting pupillary diameter. The Y type produce a rapid but short-lasting, transient response to retinal stimulation. Rapid termination of their response may be a result of feedback inhibition from the amacrine cells. Output from the Y-type ganglion cells goes not only to the lateral geniculate nucleus (LGN) but also forms a major input to the superior colliculus which controls the extraocular muscles. Their output is thought to provide the information for eye movements to locate novel or moving objects in the visual field. The X type produce a more sustained response to stimulation of their receptive fields. Their output goes mainly to the LGN and are probably more important for pattern analysis. The receptive field of the ganglion cell may be organized to enhance spatial and/or colour contrast, these features being determined by surround inhibition such as that mediated via the horizontal cell. For spatial enhancement the receptive field is arranged in concentric rings, stimulation in the different annuli producing opposite effects on the discharge frequency with respect to the cell's spontaneous level. Those cells which increase their discharge frequency in response to a spot of light in the centre of the receptive field while being inhibited by a spot of light shone on the surrounding outer parts of the receptive field have been referred to as on-centre, off-surround arrangement, while those with an inhibited central and excited peripheral field are called off-centre, on-surround types (Fig. 3.7). Colour contrast is adjusted by opposite effects on discharge frequency from stimulation by pairs of colours across the receptive field. This is known as opponent colour processing. Colour pairs which have been shown to interact in this way include red/green, blue/green and blue/yellow. These types of receptive field are found in even more developed and sophisticated forms in the LGN. Thus retinal mechanisms make important contributions to visual processing, in particular to aspects of intensity, colour, spatial and temporal analysis of our visual fields.

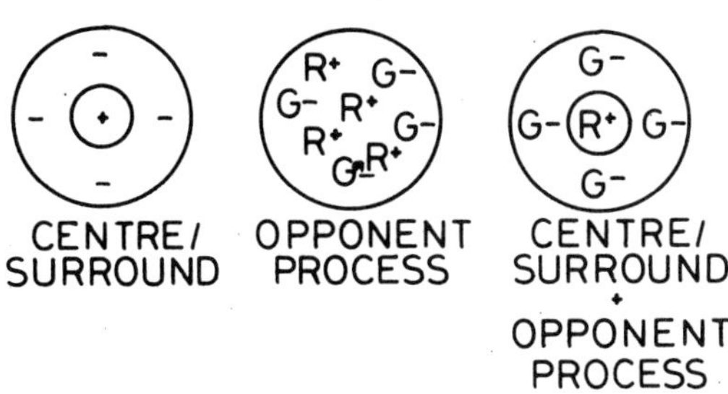

Fig. 3.7 — Receptive field types of ganglion cells. In the examples shown, R, red; G, green; +, excitation; and −, inhibition. The reverse arrangements are also seen (e.g. R−, G+; −centre, + surround) as is similar antagonism between other colour pairs.

3.7 AFFERENT PATHWAY

The optic nerves arising from the nasal half of the visual field pass to the contralateral LGN, whereas fibres from the temporal half of each field remain ipsilateral (Fig. 3.8). Collateral branches are given off en route to the LGN which terminate in the

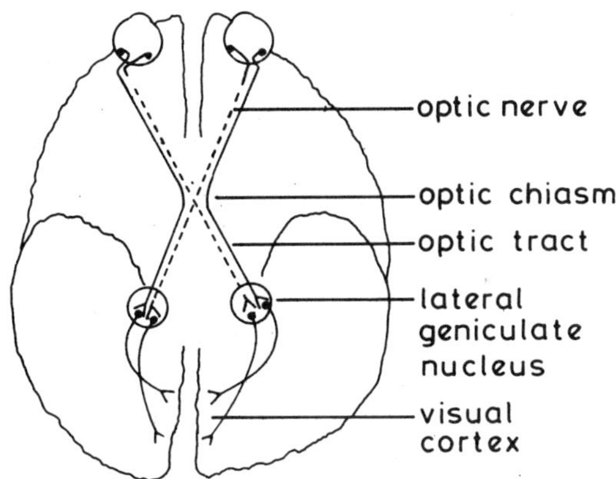

Fig. 3.8 — Simplified structure of visual afferent pathway. Branches to superior colliculi are not shown.

superior colliculus. Within the concentrically laminated structure of the LGN, fibres from the contralateral and ipsilateral eyes project in a retinotopic manner onto alternate layers. It appears, however, that the mixing of the slightly different views from each eye of any object which provides an important element in stereoscopic vision, takes place at the cortical level. In addition the LGN relay neurons are under centrifugal control from the cerebral cortex, which presumably allows selective attention to particular parts of the visual field.

The LGN neurons project to the primary visual cortex, otherwise known as the striate cortex. Anatomically this is Brodmann's area 17 located in the occipital lobe of the cerebral cortex. Detailed description of visual processing here, and in the closely associated Brodmann's areas 18 and 19, is beyond the scope of this text. However, it may be said that, largely as a result of work by Hubel and Wiesel, a hierarchical system of pattern-recognition theory was developed. In this, a succession of cortical neuronal types was described, with increasingly complex receptive fields, and it has been postulated that the receptive fields of each type were constructed from the inputs of aggregations of neurons of the previous kind. Thus the field type of a striate cortical cell known as the simple cell, which has an elongated cigar-shaped receptive field, was considered to be formed by input from an aggregation of LGN relay neurons whose input came from a linear array of overlapping retinal ganglion cell receptive fields. More complex feature extraction was then built from aggregations of simple cell output and so on up through several hierarchies. Extension of these ideas leads to the 'Granny cell' concept in which patterns as complex as a particular human face would result in the activation of a particular high-order cell. Indeed, recent evidence would suggest that there are high-order cells tuned to particular faces. However, while such studies elegantly describe the characteristics of neurons within the constraints of the necessarily simple stimuli employed, caution must be applied in attempts to relate these to the feature-extraction process. If the system worked by building complex receptive fields from simpler location specific ones, a different 'Granny cell' would be required for each location in the visual field.

Modern evidence would tend to refute the concept of such a serially organized hierarchy, offering instead a model in which there is simultaneous, parallel processing of various aspects of the visual signal by different parts of the cerebral cortex. However, this kind of approach poses problems of the unification of activity from the spatially separated groups of neurons working on stimulus location, shape, colour, surface texture, velocity, etc. How are all these facets brought together and recognized as belonging to a single object?

There is evidence to suggest that widely separated groups of neurons processing different aspects of the visual stimulus become phase-locked in their excitability levels and discharges. One way of achieving this would be by use of a neural oscillator which sends slow, regular waves of excitation through the visual cortex at a particular frequency. The various neuronal groups working on different parts of the visual signal would become phase-locked onto this wave and discharge in unison. Such synchronous discharge might carry the message that all these discharges relate to various facets of the same object.

4

Auditory system

Sound results from rapid, localized movements of molecules relative to neighbouring molecules (Fig. 4.1). For example the human voice involves the action of several

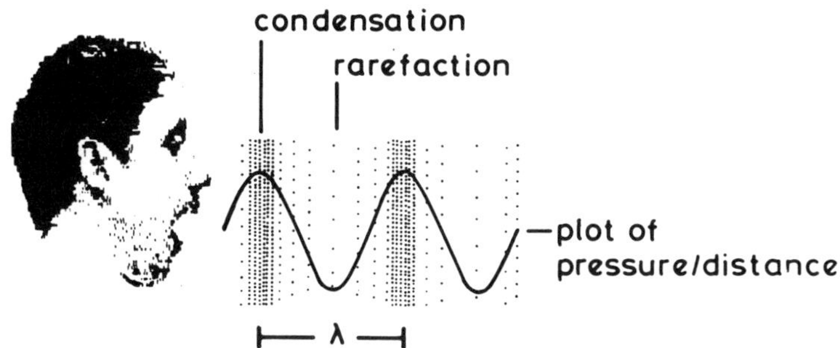

Fig. 4.1 — Propagation of sound wave through air involving cyclic changes in air pressure. For simplification, only a single, sinusoidally modulating frequency is shown here. The human voice consists of a variable mixture of many frequencies.

groups of muscles to provide a rapid flow of air and chop it up into different sounds. A simpler example is the cone of a loudspeaker, moving outwards and inwards and producing alternating localized areas of higher pressure (condensation) and lower pressure (rarefaction) of the air, which propagate outwards from the initial disturbance. These pressure fluctuations are, however, small compared with total atmospheric pressure.

4.1 PROPERTIES OF SOUND

The velocity at which sound propagates depends on the temperature and composition of the medium through which it is travelling. In air at atmospheric pressure and temperature of 20°C its velocity is around $343\,\text{ms}^{-1}$, while in water of the same temperature it travels approximately four times more quickly. Two important perceptual qualities of sound are its loudness and its pitch. These perceptions depend on the sound properties of intensity and frequency respectively, which in turn are related to the magnitude and rate of production of the pressure fluctuations. Since the range of sound intensities to which we are subjected is very wide, a logarithmic scale of intensity is used. In practice it is usually expressed as a ratio, relative to the hearing intensity threshold under ideal conditions ($2\times10^{-5}\,\text{Nm}^{-2}$). Typical intensities of some commonly encountered sounds are shown in Fig. 4.2.

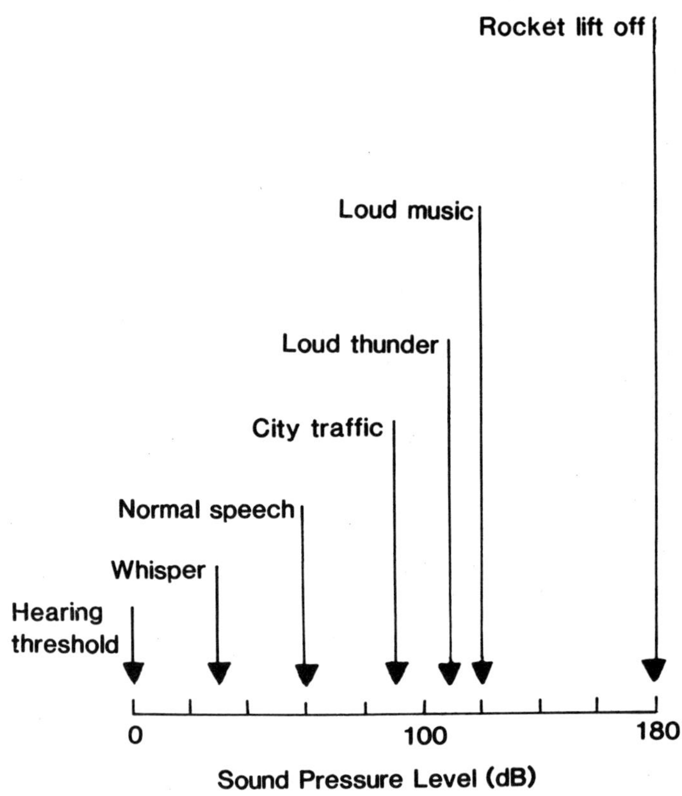

Fig. 4.2 — Intensities of some commonly (and some less commonly) encountered sounds.

The frequency range of the human ear is from 40 to $20\,000\,\text{Hz}$, with a peak sensitivity between 1 and $4\,\text{KHz}$, the upper limit declining with age. Sounds of less than $40\,\text{Hz}$ tend to be felt as vibrations. It has been said that one of the reasons a cat lies flat with its ventral surface pressed against the floor during stalking is that the abdominal skin, rich in mechanoreceptors, picks up the tiny floor-conducted vibrations produced by locomotion of small animals.

The ability to hear should not be considered simply as a pleasant, passive sensation, but also as important in generating reflexes and feedback for the control of speech. One of the commonest sounds to which we are subjected is that of our own voice, speech being perhaps the most complex motor function of which we are capable. It should therefore be of no surprise to learn that, just as various types of sensory feedback are used in the control of other types of motor activity, auditory feedback is used in the control of speech. Functionally and anatomically, the ear can be divided into outer, middle and inner portions. Let us consider them in that order (Fig. 4.3).

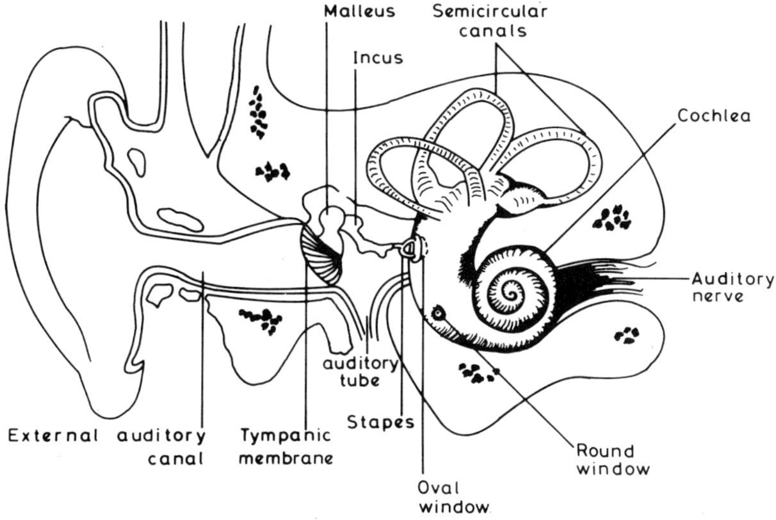

Fig. 4.3 — Structure of ear. Note division into outer, middle and inner components. The inner ear also contains the apparatus of another sensory system. These are the semicircular canals discussed in Chapter 8. Only the cochlear, auditory portion will be considered here.

4.2 OUTER EAR

First contact with external airborne sound is made by the outer ear, which consists of the flap on the side of the head — the pinna — and the external auditory meatus, which in humans is approximately 25 mm in length. The curious collection of curves and bumps on the pinna are not simply for decoration but are of importance in sound localization. Associated with the pinna are some small muscles, to allow it to be moved. One of these is the postauricular muscle. In some species, e.g. some dog breeds, these are well-developed and movements of the pinna aid localization. In humans these muscles are residual. Most people who profess to be able to wiggle their ears are actually using their scalp muscles. However, although weak and puny, the human post auricular muscle is of clinical importance, since it does reflexly contract when we hear a sudden sound. The tiny electrical impulses associated with this contraction can be recorded from the skin surface just behind the ear and may be used as an objective test of the integrity of a portion of the auditory system (Fig. 4.4).

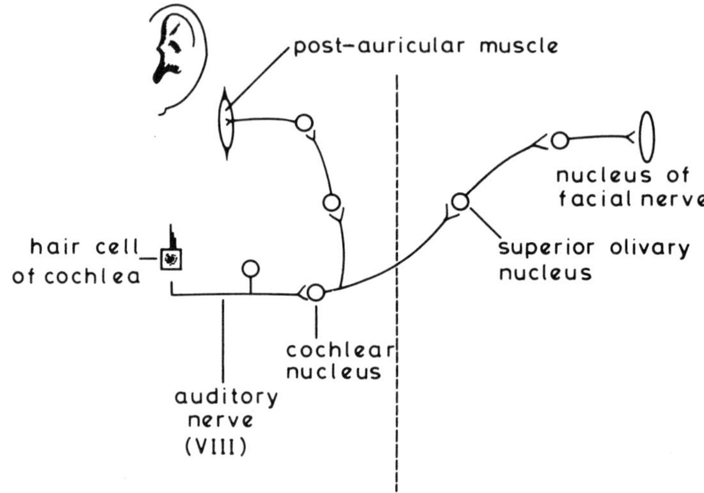

Fig. 4.4 — Afferent and efferent pathways used for the generation of the postauricular reflex. The dotted line is the midline of the neural axis.

4.3 MIDDLE EAR

The middle ear provides the link between the atmosphere and the fluid-filled inner ear (Fig. 4.5). It is limited laterally by the tympanic membrane and medially by the oval window, a membrane which looks into the inner ear. The tympanic membrane possesses differential frequency sensitivity, the portion known as the pars flaccida preferentially responding to low frequencies, while the pars tensa handles high frequencies.

4.3.1 Ossicular chain

Linking the membranes of the tympanum and oval window is a series of three, articulating bony ossicles known as the malleus, incus and stapes, which transmit sound induced movements of the tympanic membrane to the oval window. The ease with which vibrations are transmitted is controlled by two tiny muscles, the tensor tympani and stapedius, one at either end of the system. Contraction of these muscles pulls on the malleus and stapes respectively, increasing the stiffness, particularly of the pars flaccida of the tympanic membrane, and therefore decreasing transmission of low frequencies. Presentation of loud sounds (above 80 dB) produces reflex

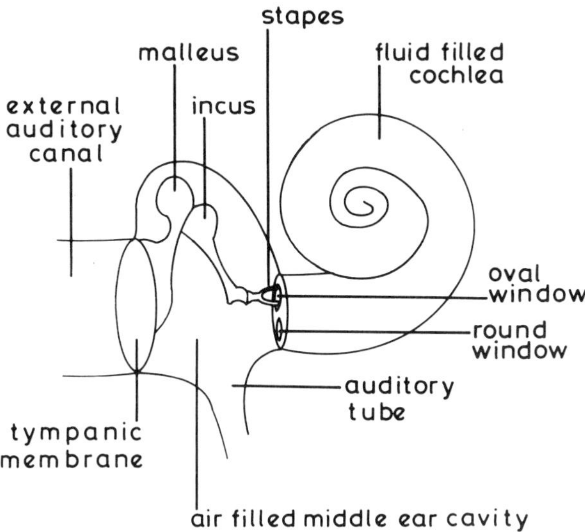

Fig. 4.5 — Diagram of air-filled middle ear to show how it links the outer and inner parts of the ear. Note the opening of the auditory tube. The other end of this tube opens into the nasopharynx, connecting the middle ear with the external atmosphere.

contraction of these muscles, although its long latency (30–40 ms) would make one doubt its action as primarily a protective reflex. It would be of no use for sudden loud sounds but would afford some protection in a continuously noisy environment.

4.3.2 Critical damping
Vibrations tend to long outlast the stimulus which initiated them. For example, a tuning fork continues to vibrate for several seconds after being struck. If this occurred in the middle ear there would be considerable interference in transmission of sound sequences. The action of the muscles, together with the elastic properties of the tympanic membrane and the ligaments around the joints of the ossicles, cause such movement to be rapidly terminated, a feature known as critical damping.

4.3.3 Pressure equalization
The tympanic membrane responds to changes in air pressure and transmits these to the inner ear. However, atmospheric pressure is not constant, changing with meteorological influences. Clearly if the air pressure in the atmosphere and that in the middle ear were different, the tympanic membrane would be deflected and its ability to respond to the tiny sound-induced pressure changes would be attenuated. There is a communicating channel between the middle ear and the external atmosphere — the auditory tube — which opens into the nasopharynx and allows equalization of pressure on either side of the tympanic membrane. Continuous gulping and swallowing of air during a rapid descent is not just a nervous reaction but a way of matching middle ear pressure with external atmospheric pressure by forcing an air bolus up the auditory tube. This tube can become blocked during upper respiratory tract infections, impairing hearing.

4.3.4 Impedance matching

The middle ear provides the link between the gas-mediated propagation in the
atmosphere and the fluid mediated conduction in the inner ear. In the condition
known as otosclerosis, parts of the middle ear ossicular chain become fused and
unable to transmit sound, resulting in deafness. Thus the ossicular chain is essential
for effective hearing. Why, then, is such a link so important? It is because different
substances offer different resistances (or more accurately different impedances,
since sound is an alternating waveform) to the passage of sound. When, for instance,
a sound passes directly from air to water, more than 99% is reflected from the fluid
surface, because in a low-density medium, such as air, sound transmission involves
relatively large-amplitude molecular excursions requiring a small amount of energy.
Transmission through a high-density medium, such as water, however, requires
relatively small-amplitude molecular excursions but larger amounts of energy to
move them (rather like trying to move a billiard ball by throwing ping pong balls at
it). What is required is a means of channeling the energy from large excursions of a
large number of molecules in the air to produce small excursions of the molecules in
the fluid of the inner ear. This is known as impedance matching. The relative area of
the tympanic membrane to the oval window, (17:1) together with a leverage effect of
the ossicular chain ($\times 1.3$) produce a concentrating factor of $\times 22$ turning the high-
amplitude, low-energy movements of the tympanic membrane into high-energy,
low-amplitude vibrations of the oval window.

4.4 INNER EAR

The inner ear consists of fluid-filled bony tubes which contain the receptors for two
different sensory systems. In the semicircular canals and the utricle and saccule are
receptors for the vestibular system, important in balance and control of eye
movements. These will be considered in Chapter 8. The cochlea contains the
receptors for the auditory system.

4.4.1 Cochlea

The cochlea consists of a fluid-filled bony tube coiled 2.75 times around a central
bony pillar (modiolus). The diameter of this tube decreases from the base to the
apex. A series of membranes run along most of the cochlear length dividing it into
three compartments (Fig. 4.6). Reissner's membrane divides off an upper (scala
vestibuli) from the middle (scala media) compartment, while the basilar membrane
separates the scala media from the lowest (scala tympani) compartment. The upper
and lower compartments communicate by a small opening at the apex called the
helicotrema. The fluid in these compartments is known as perilymph and is
characterized by a high Na^+, low K^+ concentration. The scala media on the other
hand contains endolymph, having a high K^+, low Na^+ concentration. Within the
scala media is the tectorial membrane, supported on a fibrous mat called the reticular
lamina, which in turn lies across the supportive arch of Corti. The arch of Corti stands
on the basilar membrane.

4.4.2 Basilar membrane

Sound-induced movements of the oval window causes pressure changes in the
perilymph of the scala vestibuli, which are ultimately dissipated by reciprocal

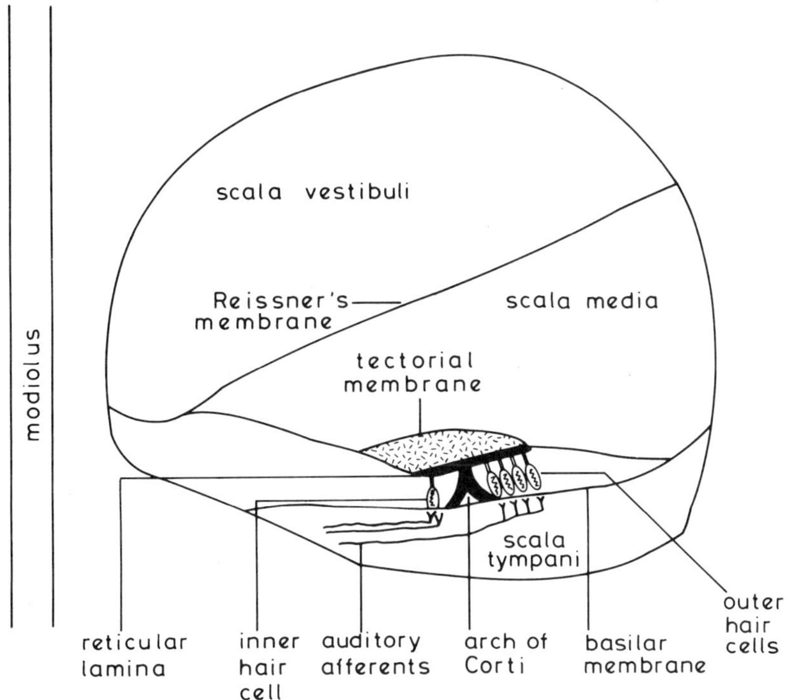

Fig. 4.6 — Cross-section through cochlea. The internal structure is essentially similar all the way along except that the diameter declines from base to apex.

movements of the round window (Fig. 4.7). To reach the round window the pressure wave is transmitted both across the basilar membrane, and via the helicotrema at the apex. The basilar membrane, approximately 30 mm in length in humans, is composed of radially arranged collagenic and elastic fibres and increases gradually in width from 0.04 mm at the base to 0.5 mm at the apex. This increase in width is accompanied by a decrease in stiffness. Hermann Helmholtz in the nineteenth century was the first to speculate that such a system could be used to separate the various frequency components in a sound. Subsequent experiments, such as those by George von Bekesy using stroboscopic light, have shown that such a spatial separation of frequencies does occur along the basilar membrane, but not by the mechanisms envisaged by Helmholtz. The pressure change in the perilymph caused by movement of the oval window produces displacement of the basilar membrane at its basal origin, setting up a travelling wave in the membrane. The interplay of increasing width and decreasing stiffness means that the lower the frequency component the further it travels along the membrane. In addition, lower frequency waves require less energy to overcome the inertia of the fluid and so travel further through the perilymph and again cause membrane displacement further from the base. The net result of interaction of the sound frequency with membrane properties and fluid inertia leads to a travelling wave in the membrane of gradually increasing

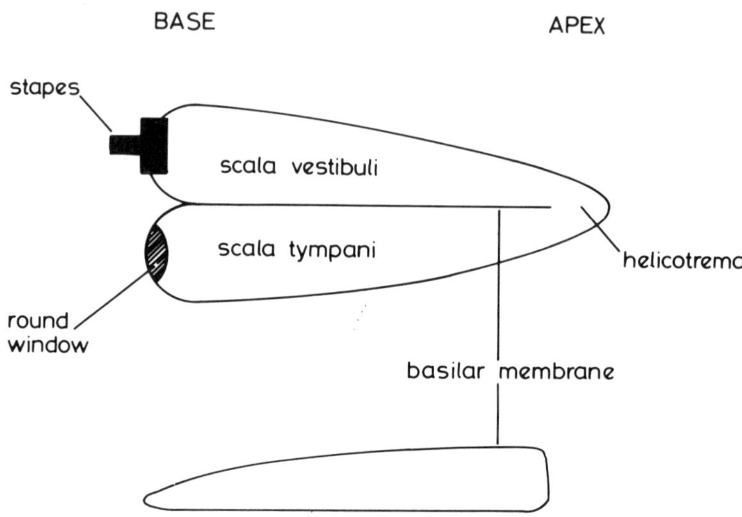

Fig. 4.7 — Diagrams of 'uncoiled' cochlea (upper) and aerial view of basilar membrane (lower). Note how the basilar membrane increases in width from base to apex.

amplitude followed by a rather more rapid decline. A spatial separation of frequencies is achieved: the lower the frequency, the further along the membrane towards the apex does it produce its maximum displacement. Thus lower frequencies will have to travel through the points on the membrane processing higher frequencies to reach their own point of maximum displacement. Similar frequencies will tend to maximally displace the basilar membrane at adjacent points. In each of these cases there will be physical interference of the processing between the two frequencies, particularly that of the higher by the lower frequency. The phenomenon is known as masking and can be demonstrated in the laboratory by the raising of the audibility intensity threshold of one frequency by the presence of a second frequency. The phenomenon of masking has also been used to advantage in the investigation of stammering to reduce the effects of auditory feedback on speech.

4.4.3 Hair cells
The next stage is the encoding of the physical displacement of the basilar membrane into electrochemical changes. This transduction is achieved by means of the hair cells, so named because of the 100 or so stereocilia projecting from their upper surface through the reticular lamina and into grooves on the undersurface of the tectorial membrane.

They are found on either side of the arch of Corti. On the lateral aspect are four to five rows of outer hair cells, while on the medial aspect is a single row of inner hair cells. The many differences between the outer and inner hair cells suggest that they are responsible for different aspects of auditory processing.

Examination of the relative numbers of hair cells and the auditory afferents with which they synapse shows that the 20 000 or so outer hair cells supply approximately 900 afferent fibres while the 4000 inner hair cells supply about 17 000 auditory efferents. Thus there is considerable input divergence from the inner hair cells, convergence from the outer hair cells. In addition the outer hair cells are richly endowed with efferents from the superior olivary complex. Recent evidence suggests that inside the outer hair cell is a latticework of protein filaments which may act like a coiled spring, opposing the viscous damping effect of the surrounding fluids on membrane movement. Efferent control of the outer hair cells may be used to selectively regulate the sensitivity of the basilar membrane or particular groups of inner hair cells, dependent on the frequencies one particularly wishes to hear in the incoming auditory message (cocktail party effect?).

4.4.4 Electrical potentials
The endolymph in the scala media is 80 mv positive, the interior of the hair cell 80 mv negative with respect to the perilymph of the scala tympani, giving a maximum available receptor transmembrane potential of 160 mv. Displacement of the basilar membrane bends the hairs on the receptor cells against the undersurface of the tectorial membrane, which in turn may lead to the opening of pores in the receptor membrane. However, whether the resulting receptor potential is due to changes in ionic conductances remains to be determined.

4.4.5 Characteristic frequency
There is a synapse between the receptor cell and its auditory nerve fibre(s). Transmission of information along the auditory nerve fibres is encoded as sequences of action potentials. It is therefore of some interest to consider the relationship between the input sound and the afferent message.

Microelectrode recordings from single auditory nerve afferents receiving input from hair cells located at a particular position on the basilar membrane show a characteristic frequency (Fig. 4.8); that is, the sound input frequency to which they

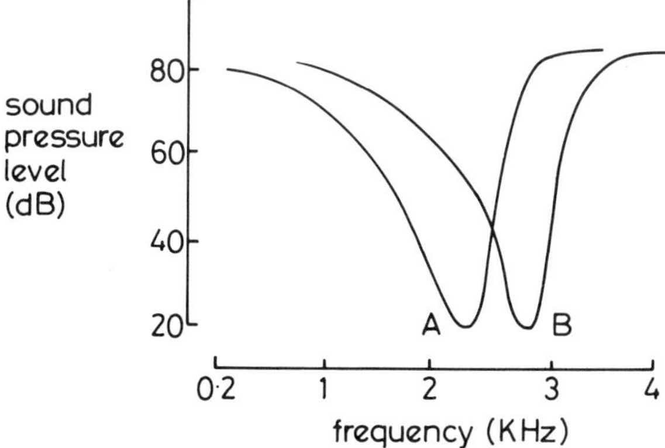

Fig. 4.8 — Tuning curves for two auditory afferents (A and B) showing their frequency selectivity. Note that the curves are not bilaterally symmetrical about their characteristic frequencies but tail off more gradually towards lower frequencies.

respond at the lowest sound intensity. From such experiments, tuning curves of frequency selectivity of the afferents can be constructed. This frequency selectivity is not surprising given the tonotopic properties of the basilar membrane. However, if one also constructs a tuning curve for the selectivity of the basilar membrane, it is seen that the afferent tuning curve, though centred on the same frequency, is considerably sharper than that of the portion of the basilar membrane which supplies it (Fig. 4.9). This seems to be yet another form of surround inhibition, in this case

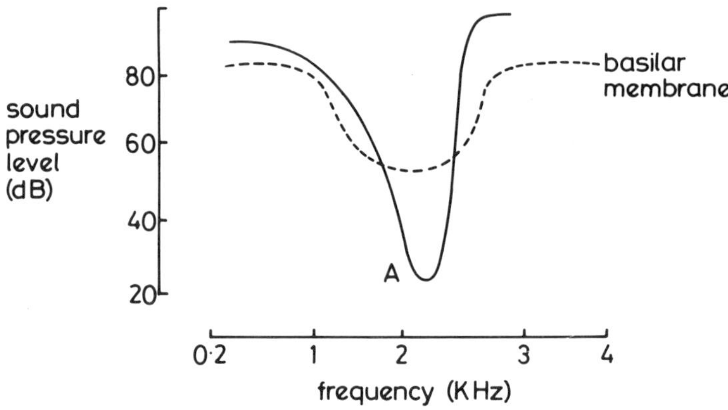

Fig. 4.9 — Tuning curves for auditory afferent A and for portion of basilar membrane served by that afferent. Note greater sharpness of tuning curve of afferent.

used to sharpen up the auditory message, operating at the level of the auditory receptors. The mechanism here is unknown. It does not utilize interneurons as is the case in the retina.

One of the mechanisms for signalling the intensity of a sound may be a characteristic frequency spread. By this is meant that as the intensity of a sound rises, so it will start to be signalled by afferents other than those for which this is the characteristic frequency. The question needs to be addressed of how the peripheral frequency analysis is signalled in the afferents and the nature of its relationship with the actual perception of pitch. A number of different theoretical approaches have been explored. The work of von Békésy, for example, lead to the development of place theories of frequency perception.

4.4.6   Place theories of afferent coding
These held that frequency perception was entirely a matter of the sorting imposed by the basilar membrane, the mechanical distortions of which would then be transduced and relayed centrally. The nature of the message was unimportant, the interpretation would depend only on where it came from.

4.4.7   Volley theories of afferent coding
While the location is undoubtedly of importance, so too is the nature of the message. For example, if we listen simultaneously to two fairly widely separated frequencies,

on the cycles in which they are in phase a third sound known as a beat can be heard. Under these circumstances there is no additional activity on the portion of the basilar membrane corresponding to this beat frequency. The perception of the presence of this frequency is not produced from its location but from the message. Alternative ways of considering frequency signalling have been developed based on the nature of the message.

In the simplest form these require that the frequency of action potentials generated in the auditory afferent should match the input sound frequency on a one-to-one basis. It can be shown that at low frequencies, nerve discharge does indeed become phase-locked onto the input frequency in this way. However, a moment's thought given to the basic properties of nerve fibres will tell us that refractoriness of around 1 ms following action-potential production will impose an upper limit of about 1 KHz on the frequency which can be signalled in such a way.

To take account of this limitation, volley theories were developed which suggested that such a signalling system could still be maintained at higher frequencies if groups of afferents worked together, each one signalling every nth wave. It is likely that all the above systems are used, the mechanism of primary importance being dependent on the frequency to be signalled.

4.5 LOCALIZATION OF SOUND

The ability to localize sound in space is a most important capacity. In the first instance the localization mechanisms trigger a number of reflexes which turn the head, eyes and attention towards the sound to allow visual examination.In order to localize sound we need to be able to estimate the distance and direction of the source. One clue we use to estimate distance are the relative amounts of high and low frequencies in the sound. Since transmission of high-frequency waves requires more energy than for low frequencies the former are more rapidly attenuated with distance. In speech, different letters contain different frequencies, vowels, for instance, containing a high proportion of low frequencies. The balance of intensities between particular letters will therefore give us an idea of the distance of their source. A number of important clues are used to analyse the source direction.

It has recently become apparent that the pinna is functional as well as decorative. Although in humans its mobility is rather limited, the curious patterns of bumps and dips on its surface provide spectral colouring, imposing notches and peaks in our frequency sensitivity curve which vary with the angle at which sound approaches the ear. For this reason, someone with only one functional ear can possess surprisingly good localization ability. However, the use of two ears allows additional mechanisms to be deployed.

Two important clues for binaural localization are present in the sound reaching our ears: its intensity and its phase. Our two ears are spatially separated by about 15 cm, with the head lying between them. If we imagine a sound source to one side then that sound passes to the near ear unimpeded. To reach the far ear, however, it must pass around the head. Some of the sound will bounce off the head rather than passing around it and therefore the sound arriving at the far ear will be of a lower intensity. Further, it is likely that sound will arrive at either ear at a different point (phase) of its cycle. This phase difference can also be measured. The relative importance of these two clues depends on the sound wavelength. If it is short as in high-frequency sound, then there is a tendency for it to be reflected from, rather than diffracted around, the head. For high-frequency sound, therefore, intensity difference is the more important clue. If the wavelength is long, however, it will easily pass around the head leading to similar intensities. However, on arrival, there will be a marked phase difference. For low-frequency sound, therefore, phase difference is the more important.

4.6 AFFERENT PATHWAY

While most of the afferent pathway goes contralateral, a smaller projection ascends on the ipsilateral side (Fig. 4.10). Further, there is a certain amount of lateralization

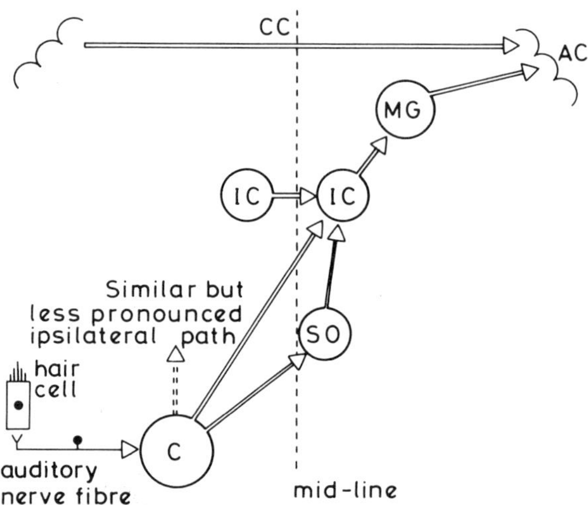

Fig. 4.10 — Auditory afferent pathway. C, cochlear nucleus, SO, superior olivary complex, I, inferior colliculus, MG, medial geniculate nucleus, AC, primary auditory cortex, CC, corpus callosum. The strong projection is contralateral, but there is also a significant ipsilateral projection.

in the auditory system in that in most people the auditory channel from the right ear is given priority in the CNS. This can be seen in dichotic listening tests in which different and complex auditory information such as word sequences are delivered simultaneously to each ear. It is found that the sequences delivered to the right ear are better recalled than those to the left. This is related to the importance of auditory feedback in the control of speech, the motor control of which in most right-handed people resides in the left cerebral hemisphere.

4.6.1 Cochlear nucleus

The auditory nerves project tonotopically onto the cochlear nucleus. Here, not only are the neurons tuned to a particular frequency, but additionally are inhibited by frequencies above and below their characteristic frequency. More complex receptive fields found here include some sensitive to the duration of the sound, others to sound containing frequency changes. Outputs from here go to the inferior colliculi, some directly, others via the superior olivary complex.

4.6.2 Superior olivary complex

The neurons in the superior olivary complex are the lowest level to receive input from both ears. We might suspect therefore that its neuronal properties might reflect some aspects of binaural processing. Each of its relay neurons possess two distinct aggregations of dendrites, one set receiving an excitatory input from one ear, the other an inhibitory input from the other ear. Some of these neurons show a characteristic delay, that is, respond optimally when there is a slight difference in phase of arrival of a sound at the ears. Others have a characteristic intensity, where optimum response depends on differential intensities at either ear. These properties would correspond with binaural localization of low- and high-frequency sounds respectively.

4.6.3 Inferior colliculus

The next link in the auditory chain is the tonotopically organized inferior colliculi, which mix spatially coded input from the superior olivary complex with the complex sound analysis from the cochlear nucleus. They seem to be important in the generation of a number of auditory reflexes, including turning the head and the attention to auditory stimuli.

4.6.4 Medial geniculate nucleus

In the medial geniculate nucleus the tonotopic organization is preserved and projected onto vertically oriented laminae. Only the ventral portion of this nucleus seems to be a specific auditory relay, the remainder being involved in complex auditory reflexes. The medial geniculate nucleus projects to the primary auditory cortex in the superior temporal gyrus.

4.6.5 Primary auditory cortex

The primary auditory cortex is in the superior gyrus of the temporal lobe. It is tonotopically organized; high frequencies most medial, low frequencies lateral, with all the neurons in a vertical column having the same characteristic frequency. However, over half of the neurons in the primary auditory cortex do not respond to single frequencies, but require presentation of complex sounds such as particular tonal sequences or durations. The auditory cortex will be considered further in Chapter 11 in its relationship with speech.

5

The tactile system

The skin has receptors for both mechanical and thermal stimuli of varying degrees of intensity. The response to noxious stimulation will be described in Chapter 7, but what of the more gentle aspects of tactile sense, such as touch, pressure, hair movement and vibration sense? Some cutaneous receptors are phasic in response, signalling the dynamic aspects of stimulation, while others give a more tonic discharge, providing information on the more static aspects of the stimulus.

5.1 RECEPTORS IN SKIN

The skin of the palmar surfaces of our fingertips is covered in tiny ridges (fingerprints). These offer a frictional surface, becoming deformed by the lightest of tactile contact which mechanically stimulates the underlying receptors. Cutaneous stimulation activates a number of histologically distinguishable types of receptor, from free nerve endings to encapsulated endings such as Meissner's and Pacinian corpuscles. Their afferents are in the Aα and A β range. Many attempts have been made to assign the various cutaneous sensations to particular receptor types. However, the extent to which this is so has yet to be determined. One problem, for example, is that some areas of skin, such as the pinna of the ear, possess only free nerve endings, yet a full range of cutaneous sensations can be evoked.

5.2 AFFERENT PATHWAY

Hemisection of the spinal cord results in the Brown–Séquard syndrome in which there is loss of fine, accurately localized touch on the same side of the body as the hemisection, but retention of a crude, poorly localized type of touch. Thus at least two ascending tracts are involved, one ascending the cord on the same side as its input, the other, which is spared in these circumstances, ascends in the spinothalamic tract of the opposite side. Further description of this latter portion will not be attempted here. The fine, accurately localised aspect is carried in the dorsal columns of the white matter of the spinal cord and has been given the snappy title of dorsal column medial lemniscal system. Its importance lies in the fact that this portion of our tactile sense is not just a passive sensing system, but is, as we shall see, directly involved in programming and activating movement during tactile exploration, manipulation or handling of objects. This tactile picture of a manipulated object is very closely integrated with the proprioceptive picture generated by the muscle spindles in the fine muscles of the hand.

5.2.1 Dorsal column nuclei
At each spinal segment, a bundle of axons from the dorsal root is applied to the dorsal column laterally to those ascending from below (Fig. 5.1). These primary

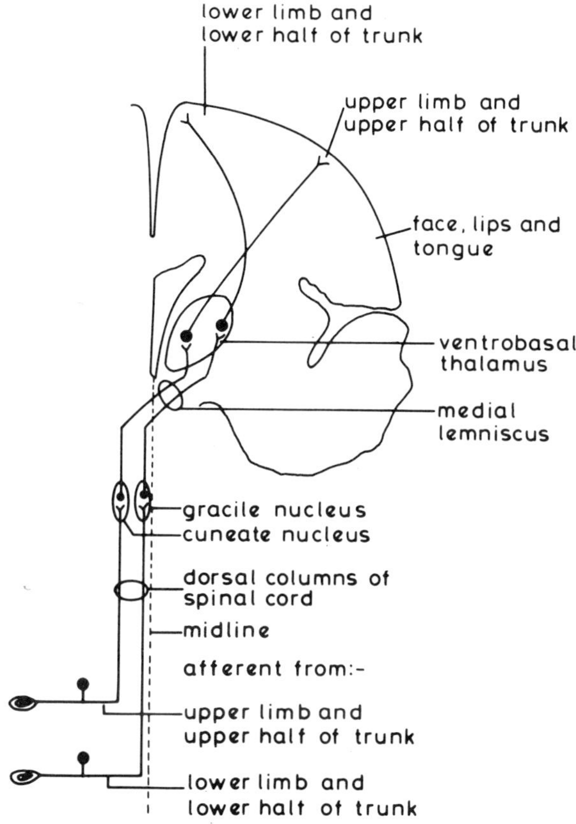

Fig. 5.1 — Dorsal column medial lemniscal pathway. Tactile sense for face lips and tongue is carried by an equivalent ascending system via cranial nerves.

fibres ascend the spinal cord ipsilaterally and synapse in the medulla, those from the hindlimb and lower half of the body in the gracile nucleus and those from the forelimb and upper half of the body in the cuneate nucleus. Equivalent input from the head and face comes via its cranial nerve to synapse in the laterally placed sensory trigeminal nucleus.

5.2.2 Ventrobasal thalamus and somatosensory cortex (sensory motor cortex)

Secondary neurons arising in the dorsal column nuclei ascend and cross the midline in fibre bundles called the medial lemniscus, to synapse in a portion of the thalamus known as the ventrobasal complex.

From here the tertiary neurons project ipsilaterally onto the primary somatosensory cortex located on the postcentral gyrus (Fig. 5.1). The topographical relation-

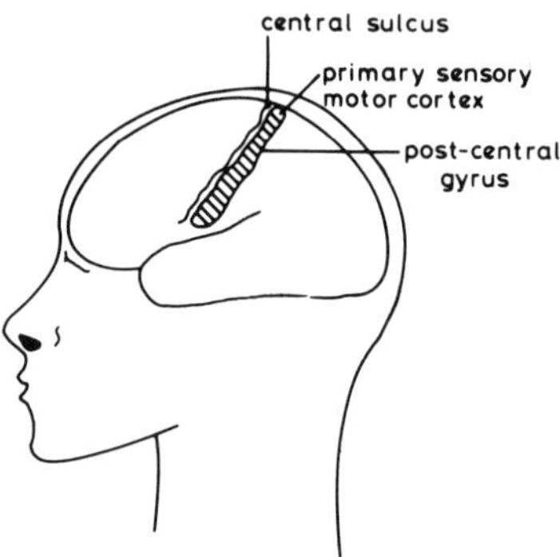

Fig. 5.2 — Location of primary sensory motor cortex in postcentral gyrus.

ships of the body are preserved in this cortical somatotopic projection, which is oriented such that caudal parts of the body are represented medially while rostral parts lie laterally.

However, there are a number of interesting discontinuities in this simple picture, some of which are produced by translating the three-dimensional bodily representation within the ventrobasal thalamus into a two-dimensional projection onto the cerebral cortex. The representation is distorted in a further, important way and is rather similar to the one displayed by the primary motor sensory cortex with which it has close anatomical and functional interconnections. The area of somatosensory cortex area devoted to the representation of a particular body part is proportional not to the surface area of that part but rather to its tactile sensitivity and motor usage. A phylogenetic trend can be observed where the extent of representation in a particular species can be correlated with its tactile behavioural adaptations. In humans, dextrous, speaking animals, large areas are devoted to digits, and intra- and perioral regions. In the rat, the vibrissal whiskers, which act as important tactile sensors, have enormous representation. Indeed each whisker can be shown to project to its own cortical cluster, or barrel, of cells.

5.2.3 Effects of dorsal column lesions

A number of lines of evidence have pointed to this system not being a purely passive input but being actively engaged in motor control. In the 1970s Dubrovsky trained cats to jump and seize a piece of meat from a slowly revolving wheel. This sequence of actions requires considerable motor planning including accurate timing of the jump to encounter the meat, precise forelimb projection, visual tracking of the object, and accurate landing. Upon recovery from surgical severance of the dorsal columns the cats showed a number of deficits, which, unexpectedly, were not primarily of tactile sensitivity, but rather of movement. Thus the cats had difficulty in initiating motor acts, and delays and inaccuracies were observed at all points in the manoeuvre. This kind of evidence led other authors to speculate that this portion of the tactile system was concerned with the planning of motor acts and with supplying the motor system with accurate tactile feedback of the evolving movement. The more complex the movement, the greater the requirement for feedback. Thus movements requiring complex, sequential motor acts would be most affected by lesions.

5.2.4   Interaction with the motor system

The structure and connectivities of this system are admirably suited for supplying continuously updated tactile information to the motor side of the CNS. The primary somatosensory cortex in the postcentral gyrus is thought to maintain a close functional relationship with the primary motor sensory cortex in the precentral gyrus by means of the long U-shaped fibres which pass under the fissure of Rolando. Such is the intimacy between the two areas that they are often considered as a single functional unit. Human tactile acuity is sharper and more able to differentiate small changes in surface texture if the fingers are actively moved over the surface than if the surface is drawn across a passive finger. Further, because of the phasic nature of many of its receptors, the sensitivity to the dynamic aspects of a stimulus, when it changes rapidly in location or intensity, is greater than that to the static aspects.

Evidence of a more direct nature can also be adduced concerning the intimate relationship between tactile input and motor output. Microelectrode stimulation of a pyramidal neuron in the primary motor sensory cortex leads to contraction of a particular muscle. If the microelectrode is now used for recording, the very same neuron can be made to discharge to a peripheral tactile stimulation. Thus tactile stimulation can directly activate the upper motoneuron and result in muscle contraction. The receptive field of the the pyramidal cell is an area of skin which would be brought into further intimate contact with the tactile stimulus upon contraction of that muscle. The fingers and thumb will tend to close around the object being manipulated, seeking out features of high contrast such as corners and edges. It has been suggested that the grasp reflex of a baby to an object placed in its hand is driven by this reflex in its raw unrefined state. In adults this system can be used for tactile guidance and the fine control of exploratory and manipulative actions of the hands. The motor output goes not only ultimately to the lower motoneuron, but also, via collaterals, into the nuclei which are sending the incoming tactile messages. Thus the motor output influences, as well as being influenced by, tactile input.

6

Taste and smell

The two sensations of smell and taste are intimately related. While smell may often be used on its own, we rely for the subtlety of our taste on a complex interaction between both smell and taste. Holding the nose when taking unpleasant medicines by mouth reduces the participation of the olfactory mucosa.

Smell is one of the most primitive of senses. Lower down the evolutionary tree, in fish and amphibia, the olfactory portion of the brain predominates. In humans, however, its representation has been reduced to the hippocampal and cingulate gyri of the limbic system. The memories which smells evoke have a powerful emotional context much exploited by manufacturers of products as varied as perfumes and washing powders. We are all subjectively aware of the links between smell, taste and the autonomic nervous system. The smell or taste of food triggers the various reflexes which prepare our gastrointestinal tract to receive food.

6.1 RECEPTORS

Taste requires that substances, dissolved or suspended in saliva are brought into contact with the collections of receptors known as taste buds (Fig. 6.1). Smell, on the other hand needs the presence of airborne odorant molecules at a concentration of 10^7 to 10^{19} molecules per millilitre of air coming into contact with the olfactory mucosa. The taste buds, found mainly on the tongue, consist of groups of 5 to 18 hair cells each possessing microvilli projecting into the mucus-filled pore of the bud. There are around 10 000 taste buds in total, which are found mainly on small elevations called the fungiform and vallate papillae.

The olfactory epithelium is some 4 to 6 cm^2 in area and innervated by around 8×10^7 free nerve endings. These are clumped together in groups of approximately 10 and embedded in supporting structures which also contain the fluid-producing serous cells (Fig 6.2). The receptive area is greatly increased by the presence of surface hairs.

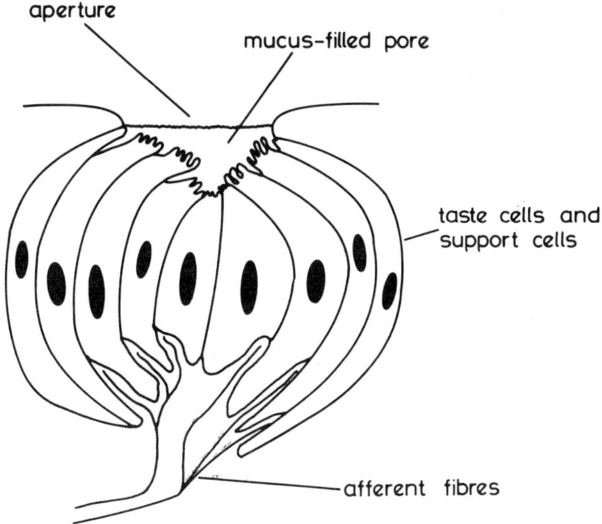

Fig. 6.1 — A taste bud.

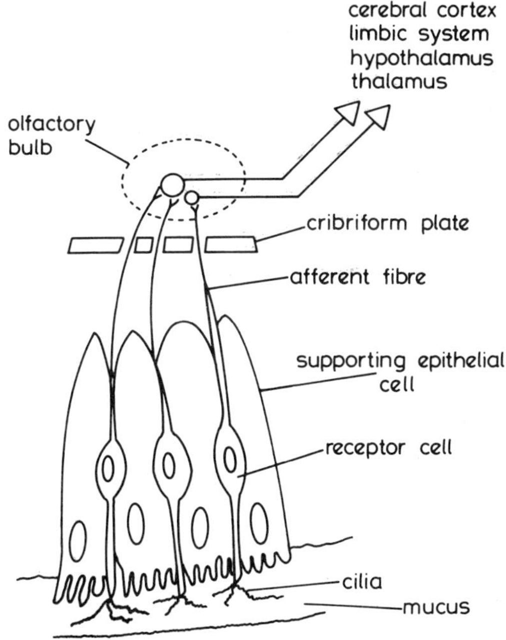

Fig. 6.2 — Olfactory receptors.

6.2 AFFERENT PATHWAYS

The anterior two-thirds of the tongue is innervated by the chorda tympani branch of
the facial nerve while the posterior one-third is served by the glossopharyngeal nerve
(Fig. 6.3). From the nasal mucosa the afferent fibres of the primary neurons pass

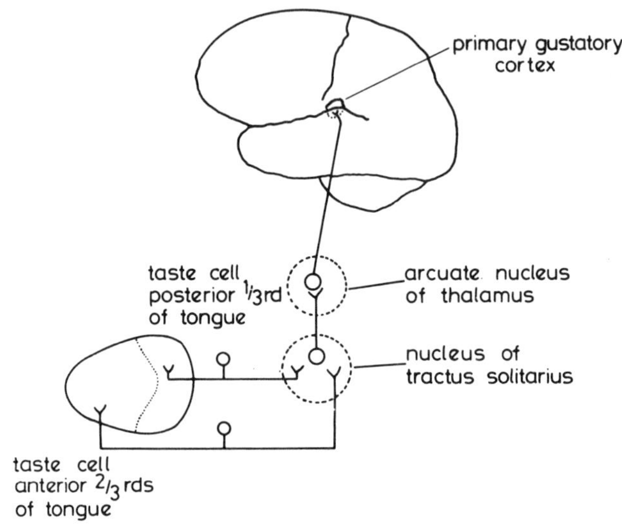

Fig. 6.3 — Gustatory afferent pathway. Note close proximity of primary gustatory cortex to the
tongue area of the primary sensory motor cortex.

through the cribriform plate and to the olfactory bulbs, where there is much afferent
convergence on an olfactotopic representation (Fig. 6.4). The primary neurons from

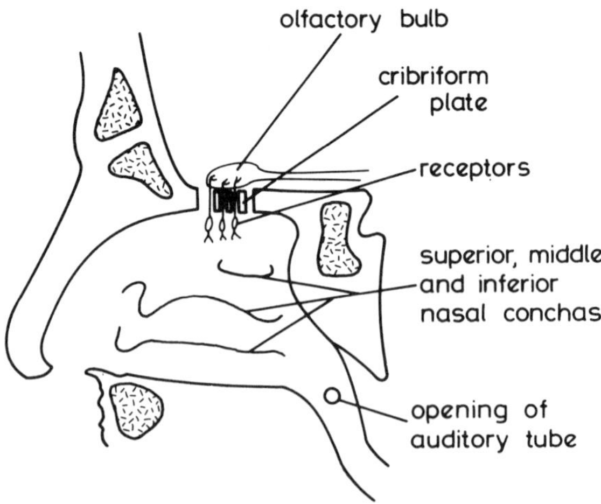

Fig. 6.4 — Location of olfactory epithelium in nasal passages and projection of receptor
afferents onto olfactory bulb.

the tongue synapse in the nucleus of the tractus solitarius. Axons from the secondary neurons join the medial lemniscus and terminate in the contralateral arcuate nucleus of the thalamus. The tertiary neurons ascend to the primary gustatory cortex on the inferior aspect of the postcentral gyrus.

The olfactory bulb neurons send inputs to many parts of the brain, including the limbic system, thalamus, hypothalamus and the uncinate gyrus of the temporal lobe. The olfactory bulbs are also innervated by many efferent fibres. In order for a substance to have an odour it must be sufficiently volatile to exert a vapour pressure which will result in the threshold concentration of molecules being reached. In practice this puts an upper limit of molecular weight on smelly things of approximately 284. There is continuous adsorption and desorption, of the odorant molecules by the olfactory epithelium.

6.3 QUALITIES OF TASTE AND SMELL

The intensity and quality of an odour depend on its molecular structure and weight, its lipid solubility and the types of chemical bonds it contains. Many classes of substances possess an odour, including halogens, sulphides, paraffins, alcohols, aldehydes, ketones, acids, benzenes, phenols, macrocyclic molecules and many others. Many attempts have been made to identify the qualities of molecules which determine their smell. One theory associates a complementary three-dimensional shape of the odorant molecule with the receptor surface protein array which would interact in a similar fashion to the lock-and-key hypothesis of enzyme substrate interaction. Another theory considers the relative amounts of infra-red emissions of molecules. However, despite many attempts, it has not been possible to relate any single pattern of chemical structure or emission spectra to odour properties.

Within a homologous series of substances the odour becomes stronger with increasing chain length, for example up to seven or eight carbon atoms, but then declines as volatility decreases. On the basis of electrical stimulation of the papillae, four basic taste sensations have been identified. These are sweet, bitter, salt and sour. They may be served by different afferent fibre sizes, since when cocaine is placed on the tongue its local anaesthetic properties cause the ability to taste bitter things to be lost first followed by salt, sour and finally sweet tastes. However, we should beware of taking too simplistic a view of taste processing. Electrophysiological recordings from afferent nerves have revealed multiple sensitivities to two, three or four of the basic tastes. A single afferent in the chorda tympani receives input from several papillae. Further, we should bear in mind that our normal 'taste' sensation will be accompanied by olfactory, tactile, thermal and kinaesthetic input.

There is a differential distribution of taste thresholds across the tongue, with sweet things being more easily tasted on the tip, bitter at the back, sour along the edge and salt along the tip and edge (Fig. 6.5). It is not clear how meaningful this simplistic classification system is in terms of taste psychophysics. Some progress has been made in relating taste to molecular configuration. Salty taste depends on the hydrated size of the cation while sour tastes depend on the presence of hydrogen ions and ease of binding of the anion. In bitter tastes such as quinine there are hydrogen bonds present with a particular spatial configuration. It is, however, difficult to specify what sweet things have in common. Sweet-tasting things include such diverse

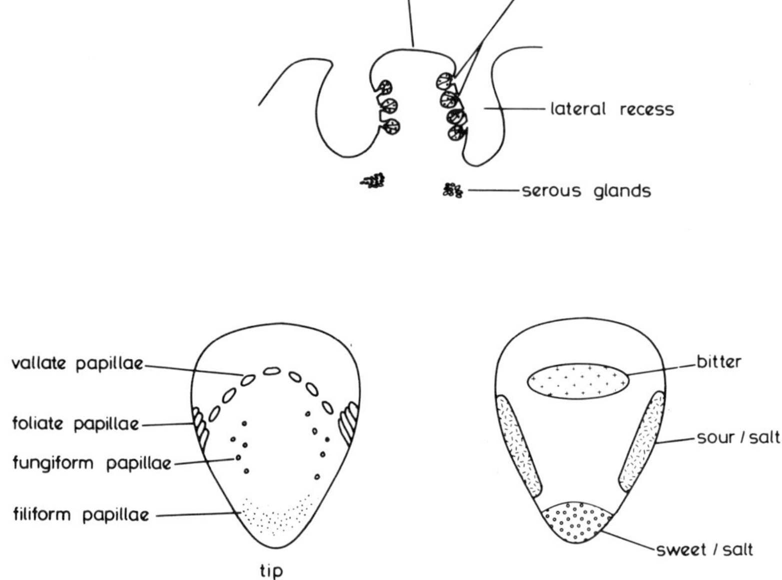

Fig. 6.5 — The upper part of the figure is a cross-section through a papilla to show the location of taste buds in lateral recess. The lower part of the figure shows differential distribution of papillae (left) and taste thresholds (right) across superior surface of tongue.

molecules as sugars, amino acids, glycerine, cyclamate and saccharine. Taste thresholds for most substances have a normal distribution through the population. However, the ability to taste the substance phenylthiourea is bimodal. We are either tasters or non-tasters. To a taster this is an extremely unpleasant bitter-tasting substance. The ability to taste it is thought to depend on the presence or otherwise of a single gene.

6.4 FUNCTIONS OF TASTE AND SMELL

Taste and smell serve a number of useful functions, including giving us sensory satisfaction and allowing us to avoid dangerous substances and seek needed substances. For example, sodium deprivation lowers the taste threshold for sodium chloride while adrenalectomy causes 'salt hunger'. Parathyroidectomy leads to a 'calcium hunger' and insulin injection increases the preference for sweet things. Thus smell and taste are involved at a very profound level in providing the drives which allow us to satisfy a number of our basic needs. This influence extends beyond various hungers and thirsts into other areas under limbic and autonomic control such as sexual drives. In many species, pheromones — types of hormones emitted by individuals of one gender — can be picked up some distance away by the olfactory or gustatory apparatus of a member of the opposite gender and act as powerful sexual stimulants. The jury is still out on the significance of pheromones in human sexual attraction. Whatever their significance to humans they seem to be species-specific. Human use of boar pheromones has had a mixed reception.

6.5 MODULATION OF TASTE AND SMELL

There is considerable modulation of our smell and taste thresholds according to gender, age and physiological status, for example to the active component of many of the musky smells in perfume. One of these is a macrocyclic substance known as exaltolide. Females have a much lower threshold to smelling this substance than do men, possibly related to trophic effects of the hormone oestrogen on the olfactory receptors. Powerful modulatory influences of steroid hormones have also been observed on the taste buds, since in males castration leads to their atrophy and involution.

7

Pain

In most circumstances, although we may not agree, pain is an extremely important sensation, signalling potential or actual tissue damage, and urging us to take corrective action to minimize that damage, for instance the rapid withdrawal of a limb from a painful stimulus by activation of the flexor reflex. The significance of pain can be illustrated by study of the tiny number of individuals born without the capacity to feel pain. In one such study by Ogden and Roberts in 1959 there was surprise that the patient was still alive at the age of 30, even though his joints were greatly damaged and deformed, half his tongue had been bitten away, his face was badly scarred and he had lost the ends of many of his fingers and toes.

An inability to feel pain leads to cumulative damage to many parts of the body, the individual being unable to adequately guard against damaging stimuli and failing to protect areas already damaged. However, there are many occasions when it would be advantageous to suppress pain, such as that associated with incurable conditions or during surgical procedures. Chronic, severe, unrelieved pain has a destructive effect on human personality and abilities. Much research has therefore been aimed at explaining and modifying the neurophysiology and neuropharmacology of pain.

7.1 SPECIFICITY THEORIES

Historically, the approach to elucidating the sensory mechanisms of pain runs parallel with the methodology employed in the study of other sensory modalities (Fig. 7.1). That is, there are attempts to identify the sensory receptors whose 'adequate stimulus' is pain, to isolate the afferent fibre groups involved in conveying the impulses, and to trace central pathways and areas of the CNS involved in processing that particular modality. This approach, while yielding valuable information, considers each sensory input in isolation, like parallel systems without interconnections. Such approaches are known as specificity theories. Let us consider how far such theories will take us in the explanation of pain processing.

7.1.1 Receptors

Iggo (1978) functionally identified three categories of cutaneous nociceptors: pure mechanical, pure thermal and polymodal. Cutaneous nociceptors have been shown

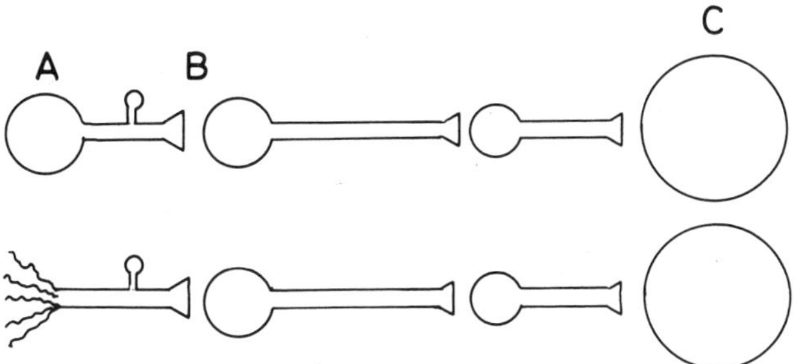

Fig. 7.1 — Diagram summarizing specificity approach to pain processing. Parallel processing systems for non-noxious (upper) and noxious afferents (lower) consisting of separate receptors (A), afferent conduction pathways (B) and central processing areas (C).

to be a particular type of free nerve ending. The threshold for activation of a mechanical nociceptor is a pressure of around 600 g cm^{-1}, while that for a thermal nociceptor is about 45°C. Pain receptors themselves are very slowly adapting. In addition, the stimulus may rupture the membranes of surrounding cells, releasing many chemical stimulants of pain nerve endings such as potassium ions, bradykinin and histamine.

7.1.2 Afferent nerves

Other groups have shown that nociceptive stimulation of the skin results in activity in two groups of afferents — A δ and C fibres. Different aspects of the pain sensation are carried in these two types of fibre (Table 7.1). The larger, myelinated Aδ fibres

Table 7.1 — Comparison of qualities of fast, slow and visceral pain

Type	Rapidity of onset	Rate of intensity rise and fall	Localization	Conducting fibres	Emotional impact
Fast pain	Rapid	Fast	Good	Aδ myelinated	Low
Slow pain	Slow	Slow	Poor	C unmyelinated	High
Visceral pain	Slow	Slow	Poor — often referred	C unmyelinated	High

carry a sharp, well-localized variety known as fast pain, which, because of the high conduction velocities of the fibres (12–30 m s^{-1}) tends to develop rapidly following stimulation.

The smaller, unmyelinated C fibres carry the more slowly developing, less well localized, slow pain, which may have an aching or burning quality. Considerable differences in the emotional impact of these two pain components can be seen, with slow pain triggering far more intense emotional responses and also autonomic reflexes, including cardiovascular and respiratory changes.

7.1.3 Central pathways

It was shown by Keele in 1957 that pain transmission up the spinal cord did not survive surgical section of the anterolateral quadrant of white matter. This part of the spinal cord contains the spinothalamic tracts, divisible into archi- and palaeospinothalamic elements carrying slow pain and an evolutionarily newer neospinothalamic portion carrying fast pain. The projections of the older parts are much more diffuse than those of the newer portions going to reticular formation, thalamus, hypothalamus, basal ganglia and frontal cortex. The neospinothalamic tract, on the other hand, goes via the thalamus to a relatively discrete area, the primary somatosensory cortex, where it projects in a point-to-point fashion.

Thus we have described a sensory system which has specific sensory receptors, afferent fibre systems, is conveyed along particular tract systems in the CNS and processed by particular areas of the CNS. The specificity approach, therefore, goes some way towards explaining the working of the system. However, there are a number of observations which are difficult to explain on this basis.

7.1.4 Problems with specificity theories

We are all aware that sensation from a painful bump can often be eased by intense rubbing or scratching of the overlying skin. Sports fields are full of people who have rubbed preparations containing capsaicin, the active principle of pepper, into their skin, to ease muscular aches. The intensity of pain perceived seems to depend on the context in which it is inflicted. To continue the sporting theme, a blow inflicted on a player during an exciting game does not have the same impact as it would if that same person was attacked on the street with equivalent ferocity.

Many other examples of the inconstant nature of pain can be found. There are, for instance, the various neuralgias which can arise as a consequence of a previous viral infection, vitamin deficiencies, diabetic degeneration or chronic alcoholism. They are characterized by an abnormal tactile sensory state in which a normally innocuous stimulus, such as gently touching the skin with a piece of cotton wool, provokes severe pain.

Following surgical amputation of a limb, a condition known as phantom limb pain often arises. In this the amputee feels that a pain-filled missing limb is still present. Pain arising from visceral elements such as the heart and gall bladder may actually be felt in some other part of the body, a phenomenon known as referred pain. A number of techniques have been developed for non-pharmacological suppression of pain. Some of these, such as dorsal column stimulation, involve electrical stimulation of other sensory systems.

7.2 SENSORY INTERACTION THEORIES

We might suspect from these examples that there are interactions between various sensory systems. The ideas which seek to explain these observations are often referred to as sensory interaction theories. A number of different mechanisms have been proposed and contradictions exist between the various models. As far as pain is concerned, the major interaction seems to be with inputs through larger fibre systems. Much of the interaction is thought to occur in the dorsal horn of the spinal cord, so it is necessary to consider the structure–function relationships here in some detail.

7.2.1 Structure–function relationships in spinal cord

The dorsal horn of grey matter of the spinal cord is divisible into six layers or laminae between and within which a number of types of interaction are thought to occur (Fig. 7.2). The cells of origin of the spinothalamic tracts are found in laminae 1, 4 and

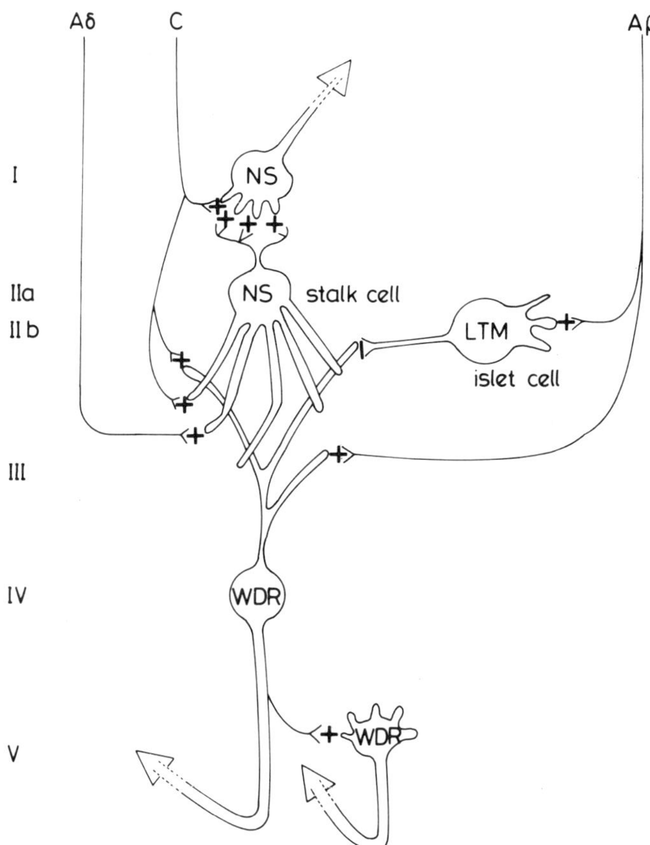

Fig. 7.2 — Neuronal interactions (much simplified) in dorsal horn of spinal cord. NS, nociceptive specific neuron; LTM, low-threshold mechanoreceptive neuron; WDR, wide dynamic range neuron. Aβ, low-threshold mechanopreceptive input; Aδ, fast pain input; C, slow pain input. (After Price, D. D. (1984) Dorsal horn mechanisms of pain. In: *Handbook of the spinal cord Vols* 2 *and* 3. Davidoff, R. A. (Ed.) Dekker, New York.

5. While those in lamina 1 seem to be very specific in being influenced only by painful input (nociceptive-specific neurons), those in laminae 4 and 5 are excited by both nociceptive and non-nociceptive inputs (wide dynamic range neurons).

Afferent access to these cells is controlled by interneurons in laminae 2 and 3, the substantia gelatinosa. Melzack and Wall likened the process to a gate which could be opened or closed to pain input. A diagram of their proposed model is shown in Fig. 7.3. It will be noted that there are a number of inconsistencies with the model

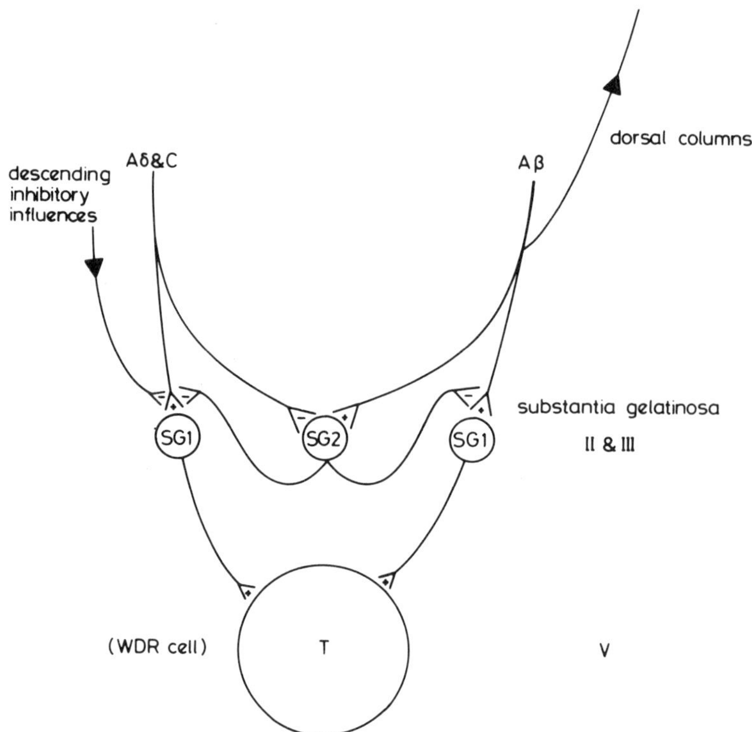

Fig. 7.3 — Segmental and descending interactions in gate theory of pain processing. (After Melzack, R. and Wall, P. D. (1965) *Science* **150** 971.)

proposed in Fig. 7.2. In particular it utilizes presynaptic inhibition to mediate the interaction. Stimulation of large A β afferents from low-threshold mechanorecepors has been shown to reduce neurotransmitter release from C-fibre terminals by presynaptic inhibition (see Chapter 2). Thus anything which increases activity in these large fibres, typically low-threshold mechanoreceptive input, will inhibit pain input.

Small interneurons in the substantia gelatinosa, known as islet cells receive excitatory input from A β fibres and turn it into a powerful inhibition of the wide dynamic range neurons. This inhibitory effect, together with the presynaptic inhibition, is relayed to adjacent segments above and below the input segment. Centrifugal inhibition of the segmental pain processing mechanism has been identified in the form of fibres descending to the spinal cord from the nucleus raphe magnus in the brainstem.

7.2.2 Pharmacological interactions

The pharmacology of these various synapses is a matter of great interest, in particular identification of the transmitters which may inhibit pain input or its processing. The neurotransmitter 5-hydroxytryptamine is involved in mediating the centrifugal inhibition, though at what point remains to be determined.

The opiate morphine has long been used for its analgesic (pain-reducing) properties. This aspect of its action is accomplished by binding with specific receptors. One of the locations of these receptors is in the parts of the dorsal horn of the spinal cord where pain processing occurs. Morphine itself, however, is not considered to be a substance produced naturally by the brain (although minute amounts of a morphine-like substance have very recently been discovered in the CNS). There has therefore been a search for the endogenous substances which combine with these receptors. Such substances were isolated from pig brains by Hughes and Kosterlitz in 1975. They were polypeptide in nature and given the name enkephalins. Subsequently, however, the morphine receptor population has been divided into a number of subtypes, each contributing to different parts of the spectrum of morphine action, analgesia being mediated particularly by the μ subtype. The enkephalins, however, do not seem to combine with this receptor type very well, but rather better with the δ subtype which mediates behavioural effects. Their precise role in pain processing is, therefore, as yet unresolved. It may be that they determine our behavioural responses to pain while other members of this polypeptide family may interfere more directly with the pain sensation.

7.3 PAIN SUPPRESSION

Consideration of inhibition of transmission in the pain system leads us naturally to discussion of the variety of methods for pain alleviation. We cannot leave this section, therefore, without mention of some of the ways of suppressing pain. Many strategies are available including psychogenic, surgical and pharmacological suppression, and neural stimulation.

7.3.1 Psychogenic suppression

Psychogenic suppression of pain is seen in applications as diverse as placebo effects and hypnosis. In the former, administration of a dummy analgesic pill, or simply the presence of a white coated 'doctor' figure, have been shown to bring symptomatic relief. Hypnotic analgesia is a well demonstrated phenomenon, allowing the patient to undergo surgery without pain. The mechanism is unknown, but the loss of voluntary control of attention may render the subject unable to select significant inputs such as pain.

7.3.2 Surgical suppression

Surgical intervention involves sectioning some discrete portion of the afferent pathways, such as the anterolateral tracts, or destroying part of the central pain processing machinery, for instance in the thalamus.

7.3.3 Pharmacological suppression

Pharmacological techniques include the use of analgesics and anaesthetics. Analgesia refers to a loss of the particular sensation of pain, whereas anaesthesia involves a loss of pain plus other sensory modalities.

Analgesics may be divided according to potency into strong (e.g. morphine) and weak (e.g. aspirin). Their modes of action differ considerably and strong analgesics are often addictive. Anaesthetics may be divided into local (localized loss of sensibility) and general (generalized loss of sensibility). One mode of action of local anaesthetics is to block the gated sodium channels in the axonal membrane, preventing the production and propagation of action potentials. A particularly important feature of the action of general anaesthetics is to markedly reduce information transmission through the various sensory relay nuclei in the thalamus. As the sensory bombardment of the cerebral cortex declines, it reverts to its more natural unconscious state.

7.3.4 Neural stimulation

Many stimulation techniques are available. It is less difficult to see possible modes of action for some of these than for others. One technique utilizes electrodes implanted over the dorsal columns of white matter of the spinal cord. These columns contain a concentration of ascending fibres derived from the afferents of low threshold mechanoreceptors. The patient carries a portable stimulator to apply current to the electrodes when the pain arises. Possible modes of action include the generation of orthodromic action potentials which activate the nucleus raphe magnus and production of antidromic impulses which invade the segmental branches of the afferents. A less invasive variation on this approach is for the stimuli to be applied directly to appropriate peripheral nerves by transcutaneous stimulation.

One of the most ancient techniques of pain relief is the use of acupuncture. In its original form needles were inserted into the skin at particular points and gently rotated by hand. In its modern form the manual rotation is sometimes replaced by electrical stimulation. It has been suggested that the needles activate large peripheral afferents thereby influencing spinal segmental processing. Unfortunately, the distribution of optimum points of needle insertion does not coincide with the segmental destination of the afferents they are presumed to activate. Thus acupuncture must work by direct activation of some central form of control, such as via the nucleus raphe magnus, or by some as yet undiscovered neurophysiological principle.

8

Proprioception and kinaesthesia

The proprioceptive and kinaesthetic functions of tactile, intramuscular and joint receptors are described in other chapters. These are of particular importance in telling us about our limb positions in the absolute sense and the positions of the limbs relative to each other. The vestibular apparatus complements this proprioceptive sense in several important ways. Using the position of the head as a reference point it provides reflex control of the extraocular muscles, which allows us to maintain the direction of our gaze in spite of any movements of the head or body. This is clearly of enormous importance for the continuation of clear visual input during movement. In addition, by its influence on our antigravity muscles, it helps us to maintain our balance.

8.1 VESTIBULAR APPARATUS

The vestibular apparatus is anatomically close to the cochlea of the inner ear (Fig. 4.3). It consists of three endolymph filled bony tubes, the semicircular canals, and two bony sacs: the utricle and saccule. The semicircular canals are set perpendicular to each other and measure rotational acceleration of the head. The utricle and saccule detect the position of the head with respect to gravity. The receptors of both the canals and the sacs are called hair cells. In the canals they are located towards one end of each tube in an expanded region known as the ampulla.

8.1.1 Receptors of semicircular canals

The hairs of the receptor cells project into a membranous sheet called a cupula, which is oriented perpendicular to the long axis of its canal rather like a sail (Fig. 8.1).

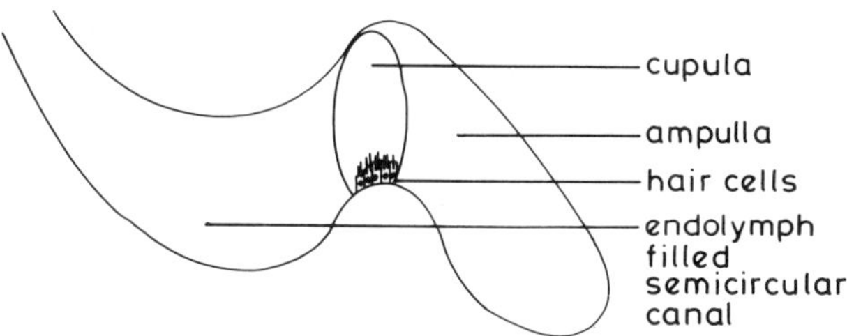

Fig. 8.1 — Cross-section through semicircular canal.

The inertia of the fluid in the semicircular canals makes its movement lag slightly behind any head movement, while the momentum gained during movement makes its cessation of motion lag a little behind that of the head. Anyone who doubts this principle has not attempted to carry a full glass of liquid across a room. In consequence, during head movement the cupula is buffeted by the endolymph, which bends the embedded hairs of the receptor cells. In the absence of stimulation the auditory afferents supplied by these receptors possess a low rate of spontaneous activity. This is modified when the hairs of the receptors are bent, being increased or decreased depending on the direction of bending imposed by the cupula. Since the inertial and momentum effects are short-lived, only movements of the head are signalled, not its static position. Thus the acceleration and deceleration associated with initiation and cessation of head movement will have opposite effects on the afferent discharge, with the relationship being reversed for head movements in the opposite direction.

Because the semicircular canals are set in different planes they will be differentially affected by any particular head movement. Analysis in the vestibular nuclei of the differential responses from the canals from the inner ear of one side and comparison of inputs from equivalent canals of both sides allows accurate assessment of the rate and direction of head movement.

8.1.2   Control of eye movement

Instructions are passed from the vestibular nuclei to the motor nuclei of the extraocular muscles to reflexly move the eyes in the opposite direction and allow the gaze to be maintained on the desired object. Its powerful effects on eye movement can be seen briefly during an experimentally induced condition called rotational nystagmus.

8.1.3   Rotational nystagmus

A volunteer is spun around rapidly for a short time in a special revolving chair. Careful observation shows that the subject's eyes keep tracking across in the direction opposite to the movement. Upon reaching the extreme limit of movement they rapidly flick back to the opposite side and start again. For about half a minute after cessation of rotation, the nystagmus continues, but in the opposite direction, the eyes now slowly tracking across the visual field in the same direction as the previous rotation, then flicking rapidly back to the other side.

8.1.4   Control of head position and balance

Vestibular output is also used to reflexly activate the neck muscles to help keep the head vertical despite changes in body posture. A third use of this information is to help maintain balance by activating both direct and indirect descending pathways to antigravity muscles. This aspect is considered further in Chapter 9.

8.1.5   Utricle and saccule

Whereas the semicircular canals signal the dynamic aspects of head posture and movement, the utricle and saccule respond to the more static elements, monitoring the position of the head with respect to gravity. Here too there is evidence for differential planar sensitivities in the utricle and saccule.

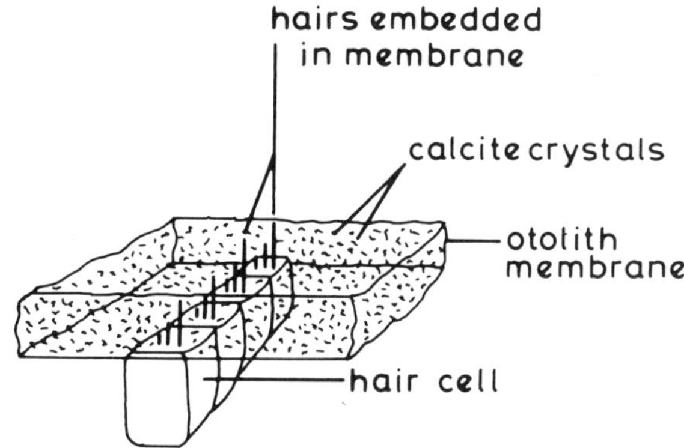

Fig. 8.2 — Receptors of utricle and saccule.

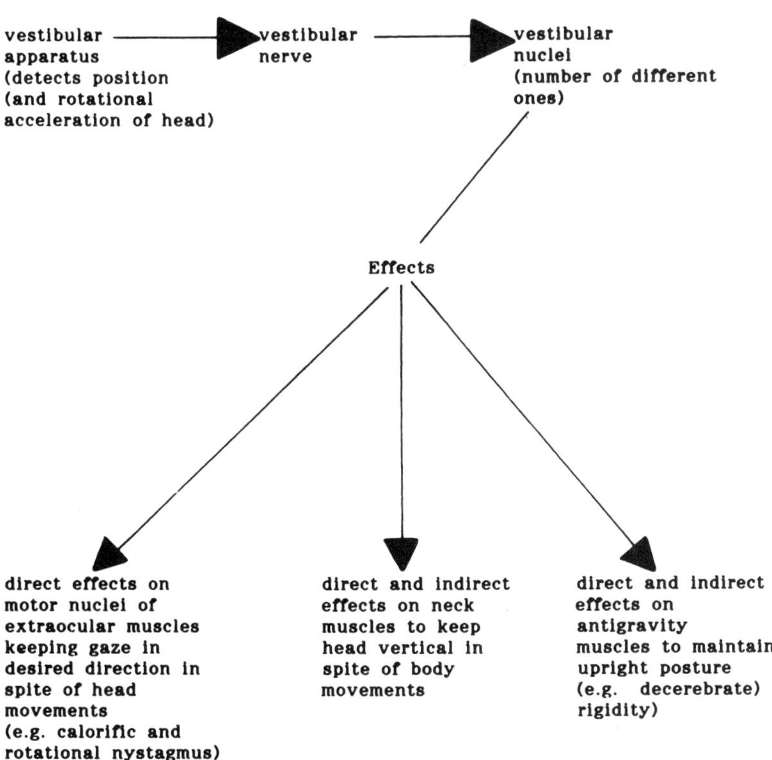

Fig. 8.3 — Summary of vestibular functions.

8.1.6   Receptors of utricle and saccule

The tufts of hairs on these receptors are embedded in a gel-like substance known as the otolith membrane, the mass of this gel being further increased by the presence of a scattering of crystals of inorganic calcium salts (Fig. 8.2). Any deviation of the head from the vertical will cause this gel to impose shearing forces on the hairs of the hair cells, bending them and altering discharge frequency in the afferents, increased by tilting in one direction, decreased by tilting in the other. The shearing force will remain constant for as long as the head remains in that position. Vestibular function is summarized in Fig. 8.3.

9

Motor control

Many parts of the central and peripheral nervous systems are concerned with the control of skeletal muscle. Particular activities require the generation of specific patterns of co-ordinated muscular contraction around a number of joints. These patterns may produce angular movement resulting in a change in posture, or resist angular movement to produce postural fixation. Both types may form part of a single action. For instance, many reaching, manipulative actions will require postural fixation of the shoulder joint with simultaneous angular movements of the digits. In addition, since the human frame is a tall structure with a narrow base, any change in the position of part of the body may require postural adjustments in other parts in order to maintain balance. Thus muscle control requires elements of movement, postural fixation and postural compensation. Groups of muscles tend to act together during movements. Those which act to move or fix the joint in a particular way are spoken of as synergists, while those which oppose them are their antagonists.

The type of movement required will determine the control mechanisms involved. The more rapid, ballistic type of movements are to a considerable extent pro-grammed centrally in advance and, once initiated, proceed with little modification by subsequent sensory feedback. In many other instances, however, the motor system does not operate in isolation from the sensory input. Indeed, the two are intimately bound. At the simplest level, sensory feedback operates to drive simple spinal reflexes, such as the rapid withdrawal of a limb from a painful stimulus. At a more complex level, sensory input supplies information to higher centres concerning the progress of a particular activity, as in the case of visual or proprioceptive movement guidance. Indeed in some instances, once the activity has been initiated, the sensory feedback may be used to actually control the movement, such as in the suckling reflex in babies upon tactile stimulation of the peri-oral region. The importance of sensory feedback in muscle control is well illustrated by the severe disruptions in speech which occur if auditory feedback is disrupted.

Contraction of a limb muscle is achieved by discharge of its motoneurons, the cell bodies of which are found in the ventral horn of spinal grey matter. These α-motoneurons are influenced by a number of descending systems, which may be divided anatomically into those which pass through the medullary pyramids — the pyramidal system — and those which do not — the extra- or parapyramidal system. The descending fibres innervate not only the α-motoneuron in the spinal cord but also a second type of motoneuron, the γ motoneuron, the functions of which will be described later.

It is clear, therefore, from the foregoing description, that motor activity is a complex phenomenon requiring contributions from many parts of the nervous system. A hierarchical system of three tiers seems to operate, incorporating not only downward, but also upward control. The lowest levels are the segmental motor programs in the spinal cord, together with simple spinal reflexes. The middle level of control involves the various projection areas to the brainstem motor nuclei and the highest levels the central programs and executive areas of the cerebral cortex, cerebellum and basal ganglia. Each level is subject to sensory feedback. Let us consider these various levels of control, starting at the lowest level.

9.1 LOWEST LEVEL OF CONTROL — SPINAL CORD AND REFLEXES

Within the anterior horn of the grey matter of the spinal cord is the α-motoneuron, the final common point of output for all the various control systems. Its axon leaves the spinal cord via the ventral root and terminates at the neuromuscular junction within the muscle innervated. Excitability of the α-motoneuron is determined by descending systems, by sensory feedback at the segmental level via the dorsal roots, and by feedback resulting from its own output. Activities as complex as locomotion may be controlled, though not initiated, to a considerable extent here by neuronal circuits known as spinal motor generators. Let us consider first a simple type of feedback and how this triggers a second, more complex response at the spinal level.

9.1.1 Flexor reflex
A sudden, painful cutaneous stimulus activates the nociceptors in the skin. Their sensory fibres, via multisynaptic spinal pathways, excite the motoneurons controlling all the flexor (joint-closing) muscles and inhibit those controlling the extensor (joint-opening) muscles in that limb, with the result that the limb is removed rapidly from the potentially damaging stimulus. This is the flexor reflex (Fig. 9.1). The conscious 'ouch' follows in due course. Suppose now that this stimulus had been delivered to one of our feet. Just as we were congratulating ourselves on having removed our foot from a nasty stimulus we would realize that we were standing unbalanced with all our weight borne on one leg and would possibly fall back onto the stimulus.

9.1.2 Crossed extensor reflex
Fortunately, this does not occur, because the noxious input which generates a flexor reflex on the ipsilateral or same side is made available to the contralateral or opposite side of the spinal cord where it generates the crossed extensor reflex. This involves activation of the gravity-resisting extensor muscles, enabling the limb in ground contact to bear the full weight of the body, accompanied by a tilting of the body to that side (Fig. 9.2).

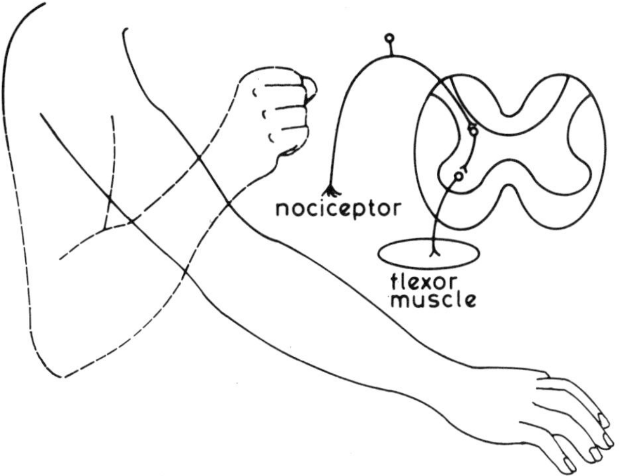

Fig. 9.1 — Flexor reflex activated when fingers encounter a noxious stimulus. Note withdrawal of whole arm and closure of joints. Pathway in spinal cord shown in highly schematized form. This reflex utilizes a variety of interneuronal types and number of synapses to gain access to the motoneurons.

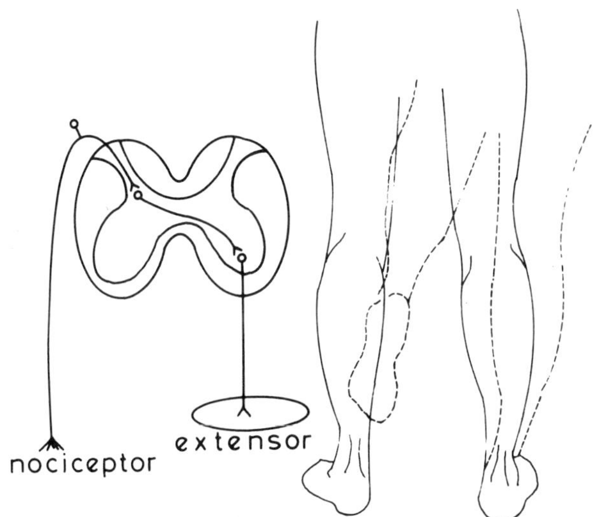

Fig. 9.2 — Crossed extensor reflex activated when foot encounters noxious stimulus and is withdrawn. Antigravity muscles of other leg are activated and body tilted to that side. Pathway in spinal cord shown in highly schematized form. This reflex utilizes a variety of interneuronal types and number of synapses to gain access to its target motoneurons.

9.1.3 Myotatic (stretch) reflex

A number of reflexes important in motor control originate in receptors in the muscle–tendon system. One of these is the stretch or myotatic reflex which can be elicited from most muscles. A specific example is the knee-jerk reflex. Striking the patellar tendon causes a stretching of the muscles attached to it. A few milliseconds later these muscles contract resulting in the famous knee-jerk. This is a monosynaptic reflex in which rapid stretching of the muscle results in activation of its own (homonymous) motoneurons, causing contraction (Fig. 9.3). The receptor involved

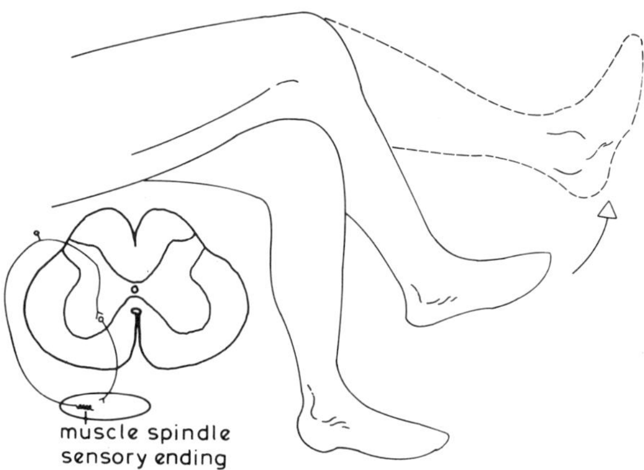

Fig. 9.3 — Myotatic reflex activated when a muscle is stretched. Reflex contraction of that muscle follows a few milliseconds later. Reflex utilizes a monosynaptic pathway in spinal cord.

seems to be detecting length changes in the muscle and is known as the muscle spindle. This receptor is of considerable importance in many aspects of motor control and is worth considering in some detail.

9.1.4 Muscle spindle

The muscle spindle is an encapsulated receptor, up to 10 mm in length and around 400 μm in diameter at its expanded midregion (Fig. 9.4). Several hundred of these

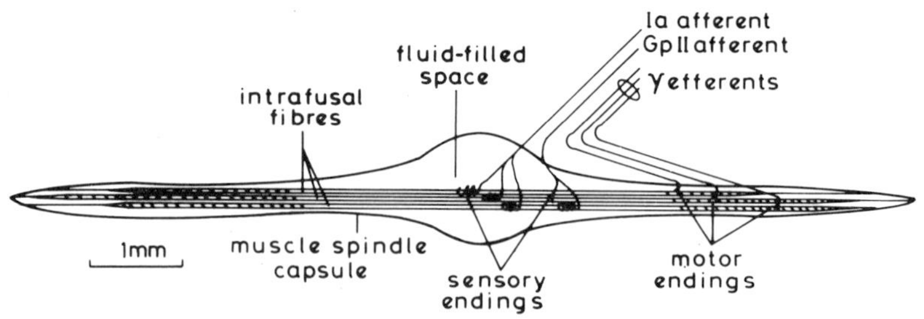

Fig. 9.4 — Structure of muscle spindle. Note striated appearance of each end of intrafusal fibres.

receptors may be found scattered throughout the muscle. The muscle spindle is mounted piggyback on a muscle fibre. Thus stretching the muscle results in stretching of the muscle spindle. Within the fluid-filled capsule are between 6 and 12 modified muscle fibres — intrafusal muscle — running the length of the spindle.There are several types of intrafusal fibre within any one spindle, which have somewhat different properties. However, those differences need not concern us here. Each intrafusal fibre, whatever type, has a non-striated central region,and it is here we find the sensory nerve endings, of which there are also several types. Stretching the intrafusal fibre stretches the sensory ending, which then develops receptor potentials. Some of these endings, the primary endings, are sensitive to both the static (length) and dynamic (rate of change of length) components of a stretch and give rise to Ia afferent fibres. Others, the secondary endings, are sensitive only to the static component of the stretch and have group II afferents (Fig. 9.5).

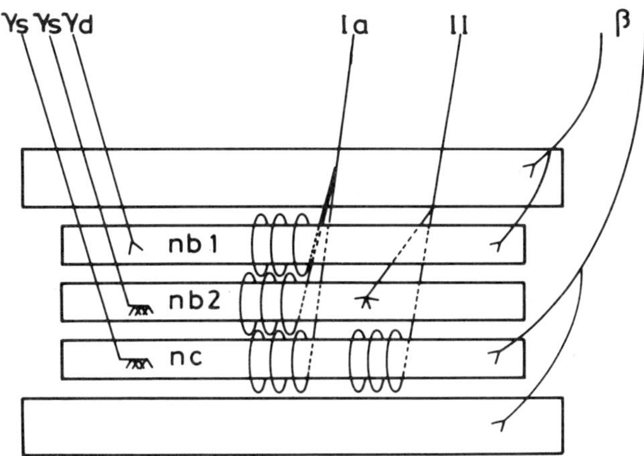

Fig. 9.5 — Distribution of sensory (Ia and II) and motor (γ_s, γ_d and β) fibres amongst the various types of intrafusal fibres (nb1 and nb2, nuclear bag types 1 and 2; nc, nuclear chain).

The demonstration of muscle spindle activity by the knee-jerk reflex is a very extreme response. It is reasonable to suppose that muscle spindles are not there simply to cause massive muscular contraction in response to someone hammering our tendons. Rather we should think of them as devices which constantly supply the CNS with information concerning muscle length. At the spinal level this information can be used to activate the homonymous motoneurons to produce length correction. One can see its uses, for instance, in muscles used to resist gravity. In humans, the lower limb extensors become stretched as the joints in the legs tend to collapse under body weight. This activates their muscle spindles, leading to reflex contraction, and maintenance of upright posture.

Again we see simple reflexes building into more complex responses. The same input is used to produce increased excitability (facilitation) of motoneurons of synergists and inhibition of those of antagonists. In addition, ascending branches make this information available to higher centres.

9.1.5 Contributions to proprioception and kinaesthesia

Recent evidence would suggest that these ascending messages make considerable contributions to our conscious appreciation of proprioception and kinaesthesia (sense of position and movement). For instance, in a human volunteer, local anaesthesia may be used to eliminate input from any nociceptive, tactile or joint receptors in the vicinity of the terminal joint of a finger. Small incisions are then made in the skin over the joint to expose the tendons of the muscles which move that joint. Pulling of those tendons by the experimenter induces in the subject a kinaesthetic illusion of bending of the finger.

Since the muscle spindle is so important for the generation of proprioceptive information when a muscle is being stretched, we might expect it to be at least as important when the muscle is contracting. However, the in parallel mechanical arrangement of the spindle within the muscle means that when the muscle contracts, the spindle will shorten, no stretching of the sensory ending will occur, and no information will be generated. To allow the spindle to generate information during contraction and to match its length to the current working length of the muscle, use is made of the contractile nature of the intrafusal fibre, which is controlled by a second type of motoneuron — the γ-motoneuron. We shall return to this phenomenon in section 9.2.1. At this point it would be convenient to outline the feedback from one other intramuscular receptor — the Golgi tendon organ.

9.1.6 Golgi tendon organ

These receptors, located at the muscle–tendon boundary, are used to monitor muscle tension. On the basis of experiments in which passive muscles were stretched, it was thought that tendon organs were sensitive only to very high tensions, their putative function being to inhibit the discharge of the homonymous motoneurons and terminate muscle contraction at high tensions, perhaps in order to avoid muscle damage. Indeed, the major consequence of their discharge appears to be inhibition of homonymous motoneuron discharge, and in some types of spasticity sudden collapse of muscle resistance to stretch can be observed. This is known as the clasp-knife reflex. However, recent work has shown that these receptors are extremely tension-sensitive during active contraction and therefore probably contribute tension information at all levels of contraction. The sensitivity of the spinal interneurons which mediate their segmental effects is under central control.

9.1.7 Joint receptors

Both phasic and tonic types of joint receptor have been identified. Early work suggested that the tonic variety had a 'best angle' that is a small angular range within the total range of a joint movement, over which it fired, each receptor signalling a different part of the range. However, more recent work has shown that these data were largely the result of the mechanics of the recording system, and that under normal conditions these receptors do not fire at intermediate joint positions. Rather, they signal extremes of joint flexion/extension, which may be of use in the generation of cyclic activities such as locomotion and in the measurement of the forces acting on a joint. It would seem, therefore, that most proprioceptive information is generated by muscle spindles. A second type of joint receptor may, however, be of proprioceptive significance during very rapid movements.

9.2 MOTOR INNERVATION OF THE MUSCLE SPINDLE

The axons of the spinal motoneurons to skeletal muscle have conduction velocities in the A α range and have therefore been referred to as α-motoneurons. In the 1930's a second group of motor fibres in the ventral root with conduction velocities in the A γ range was described. It was discovered that these γ-motoneurons innervated the striated polar regions of the intrafusal muscle fibres of the muscle spindles (Fig. 9.5). Discharges in these fibres cause the striated ends of the spindle to contract, producing profound changes in its response characteristics. Since the spindle is mechanically in parallel with an extrafusal fibre and indeed anchored to it at either end, contraction of the intrafusal ends results in a stretching of the central region of the intrafusal fibre and hence of the sensory nerve ending in this area.

9.2.1 Servo-assistance of contraction

These observations led to a radical rethink of how movements were produced. It was postulated that a muscle contraction, once initiated by descending commands, could proceed automatically to the desired end point. This would clearly confer enormous benefits in terms of the amount of intervention by higher centres required to complete a movement. Briefly, a command for movement would descend to the spinal cord where it would activate the γ-motoneuron, leading to intrafusal contraction and an elevated muscle spindle discharge. The excitatory feedback onto the α-motoneuron would lead to extrafusal muscle contraction, which would continue until the extrafusal muscle length matched the intrafusal muscle length and muscle spindle discharge returned to its previous level. Thus contraction would be controlled by a follow-up length servo.

If intrafusal contraction did indeed precede extrafusal contraction, however, then so should the elevated muscle spindle discharge which produces it. Unfortunately, under most circumstances, it does not. The theory was therefore modified to one in which the α and γ systems are activated at the same time, leading to an error signal being generated by the muscle spindle if there is any difference in amount of contraction between the two systems, with automatic correction of extrafusal contraction via the feedback loop–servo-assistance of contraction.

This may well be the way in which we accomplish small, slow, finely controlled movements such as exploratory, manipulative actions of the digits but is not of use in more rapid, gross movements.

9.2.2 Differential effects of γ-efferent activity

Functionally, two types of γ-motoneuron can be distinguished, which appear to modify different aspects of spindle function. These are known as γ dynamic and γ static fibres. The γ static fibres seem to be involved in the aspect of spindle function mentioned in the previous paragraph. On the other hand, if localized cooling is used to prevent the actions of the CNS areas which control the γ-dynamic neurons and hence the velocity sensitivity of the spindle,the most consistent deficit is that of overshoot — reaching actions which overshoot the desired goal. It seems that, during muscle stretch, velocity sensitivity may be used to impart a certain predictive capacity on the measuring system, necessary because the time lags in the system due to conduction time mean that reflex corrective action based simply on length would always underestimate the length change because in the interim further length changes would have occurred. Velocity sensitivity could be used to predict and correct for the actual length changes which would have occurred taking into account the time lags in the system.

9.3　MIDDLE LEVEL OF CONTROL — BRAINSTEM MOTOR NUCLEI

This level of control receives projections from above and below, mediates a number of important reflexes concerning posture and balance, and contains some of the nuclei which are important outputs for the extrapyramidal system. Descending fibres arising from here have actions on both α and γ types of motoneuron. It includes a number of nuclei in the brainstem such as the red nucleus, vestibular complex, inferior olivary nucleus and portions of the reticular formation. Let us consider some of the actions on motor activity. Since this level is subject to powerful control from above, to gain some idea of its activities we can observe what happens when these descending influences are removed.

9.3.1　Decerebrate rigidity

It has been known for many years that damage to certain parts of the forebrain gave rise to an increase in limb stiffness, particularly in the antigravity muscles, which allow us to reflexly maintain our upright posture in the face of gravitational forces which are tending to make us collapse in a heap. In most animals these antigravity muscles are the extensors, that is muscles which are keeping the joints open against the collapsing force of gravity. In the sloth, however, which spends much of its time hanging from trees, it is the antigravity flexor muscles which are so affected.

The investigation of antigravity responses probably gave some of the earliest clues as to the complexity of the systems governing posture and movement. The ability to produce and maintain an antigravity response is seen in an exaggerated form in the condition known as decerebrate rigidity. Charles Sherrington in 1898 described such a condition in a cat in which the brain had been surgically sectioned between the upper and lower colliculi. Later work has shown in this type of preparation that while the exaggerated antigravity stance occurs as a result of the removal of descending inhibitory forebrain influences on midbrain and hindbrain structures, EEG recording from the cortex shows the forebrain to be in a state of continuous sleep due to the lack of excitatory midbrain influences which would normally keep it awake.

The same effect on muscle rigidity is obtained if the blood flow to the forebrain is compromised by occlusion of the carotid and vertebral arteries. Either of these approaches effectively abolishes any influences from cerebral cortex, thalamus, basal ganglia and hypothalamus, while sparing those from the cerebellum and brainstem. The resultant condition seems to be occurring as a result of the actions of some of the pontine and medullary motor nuclei on antigravity responses being allowed to occur without modification from higher centres. The posture is such that the limbs are stiff and extended, the neck and tail hyperextended and the back arched.

The pyramidal tracts are not involved since a unilateral midcollicular section results in rigidity developing on the same side rather than the opposite side as would be expected to occur with a crossed pathway. Indeed it can be shown that direct activation of the spinal α-motoneuron by descending pathways is not involved at all. The most compelling evidence for this is that the rigidity collapses if the dorsal roots of the spinal cord are severed. These, you will recall, convey sensory information into the cord from the various receptor types including muscle spindles. If, in a decerebrate animal, all the antigravity muscles except one are dennervated, the single remaining muscle is found still to be stiff. Its stiffness immediately collapses upon severing the dorsal root containing its afferents.

What these pieces of evidence point to is that the rigidity seems to result from some kind of excitatory impulses coming from the muscle and reflexly exciting its own α-motoneurons. We have already seen an example of α- motoneuron activation from muscle afferents in the myotatic reflex. It seems that a similar sort of mechanism operates here. The myotatic reflex in the antigravity muscles is more easily triggered. The reason is because of an increased activation of γ-motoneurons innervating receptors in antigravity muscles. This causes contraction of the intrafusal muscle fibres within the muscle spindles in these muscles, which has the dual effect of making the sensory endings in these spindles more sensitive to any subsequent stretch and also increases the afferent discharge from these endings, leading to increased excitability of the α-motoneurons in the motoneuron pool of the muscle. Each antigravity muscle is affected similarly, with an increase in its muscle spindle stretch sensitivity, increased excitatory afferent discharge and subsequent increase in α-motoneuron excitability. We need now to determine the source of the increased γ-motoneuron activity. If, in a decerebrate animal, a section is made below the medulla, the rigidity collapses, resulting in a state of flaccid paralysis.

Thus it appears that the condition arises as a result of influences from structures in the pons and medulla, the activities of which are normally partially suppressed by descending influences from the forebrain. These structures are the brainstem motor nuclei, in particular, parts of the pontine and medullary reticular formation giving rise to the reticulospinal tracts and parts of the vestibular nuclei, notably Deiter's nucleus, by way of the lateral vestibulospinal tracts. Of the two descending pathways the former has the more powerful effect since the rigidity can be maintained by it alone.

9.4 HIGHEST LEVEL OF CONTROL

In this section I shall describe three major components: the cerebral cortex, cerebellum, and basal ganglia.

9.4.1 Cerebral cortex
In the 1870s Fritsch and Hitzig demonstrated that stimulation of an anterior area of the cerebral cortex of the dog resulted in contraction of the muscles of the

contralateral side of the body. Other workers subsequently confirmed this in primates. The area became known as the motor cortex and was considered to be the part of the brain which not only initiated movement but also contained our will to move (Fig. 9.6).

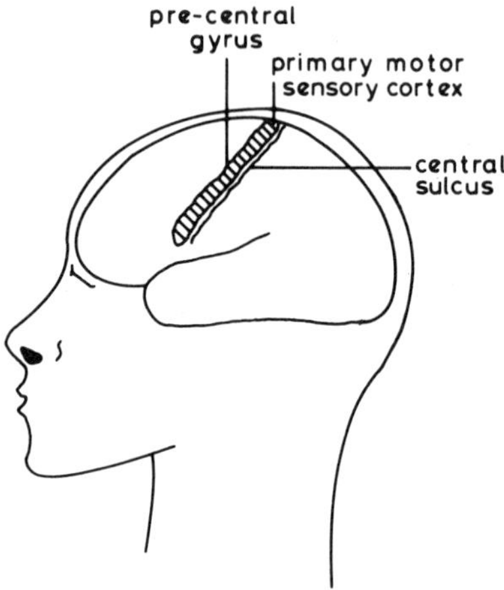

Fig. 9.6 — Location of primary motor sensory cortex in precentral gyrus.

If a subject is asked to make a willed movement, motor cortex activity precedes signs of muscle activity by around 100 ms. However, activity over wide areas of the brain outside the motor cortex can be recorded up to 800 ms prior to movement. Thus while the motor cortex can perhaps be considered an important point of final output for the motor system, activity relating to the decision to move appears to involve much wider areas of the brain. Even the term motor cortex is somewhat inaccurate since other separate discrete cortical motor areas exist. This original area, therefore, became known as the primary motor cortex. Recent work has shown that the primary motor cortex receives considerable sensory input, such as from cutaneous tactile receptors, the significance of which will be described later. It is thus more accurately referred to as the primary motor sensory cortex (Ms1). However, for the sake of brevity, I shall refer to it simply as the motor cortex.

9.4.2   Primary motor sensory cortex

Anatomically, it occupies the precentral gyrus. Stimulation of discrete areas of the motor cortex using microelectrodes results in contraction of particular muscles, and there is a consistent relationship between the position stimulated and the muscle activated. This topographic layout is shown in Fig. 9.7. The muscles of the toes and feet are represented most medially, face and tongue most laterally.

If we look now at the relative areas of motor cortex associated with control of particular muscles, we find that the muscles used for the finest, most skilled

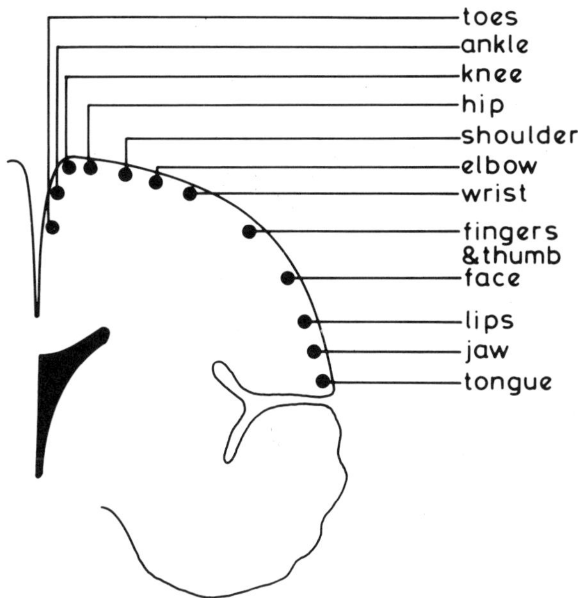

Fig. 9.7 — Topographical layout of primary motor sensory cortex.

movements, are given the largest areas, and hence the greatest number of neurons, for their control. In humans this means that muscles controlling the digits and mouth are served by large areas; in other species other muscle groups are finely controlled and have large areas devoted to them, such as the muscles controlling the prehensile tail of the spider monkey.

9.4.3 Output from motor sensory cortex

The motor cortex is a six-layered structure, the output cells being found in laminae 3 and 5. These neurons are pyramidal in shape and vary considerably in size. The largest are often known as Betz cells and were originally thought to be the sole output cells of the motor cortex. One of the routes from the pyramidal cells to the motoneuron in the spinal cord is a direct one, the fibres of which cross the mid-line in the medullary pyramids to contribute to the corticospinal tract. Since there are around 30 times as many axons in the corticospinal tracts as there are Betz cells, it is clear that the smaller pyramidal cells must also output into this system. Indeed more than 50% of the fibres seem to originate from cortical neurons outside the motor cortex. Innervation of the spinal motoneuron by this system in humans is largely contralateral, and may be mono-, di-, or trisynaptic. The greater the degree of fine control exerted over a particular muscle, the fewer the synapses. This system has also been called the pyramidal system, defined, perhaps a little confusingly, not by the cells which give rise to it, but rather by being formed from the fibres which pass through the medullary pyramids.

Descending systems also reach the spinal motoneuron by a number of indirect routes. These are often referred to collectively as the extrapyramidal or parapyramidal systems.

9.4.4   Neuronal arrangement

The neuronal arrangement within the motor cortex is of some interest. As in many other areas of cortex, the neuronal arrangement is columnar, with overlapping mosaics of columnar efferent zones to muscles around a particular joint. The representation of the different joints within a limb seem to be arranged as nesting concentric columns. For instance, within the arm area, the shoulder columns surround the elbow columns, which in turn surround the wrist columns and finger columns, rather like the skins of an onion. The question has not yet been resolved as to whether the motor cortex simply controls individual muscles, or whether it produces co-ordinated movements. Intracellular recordings from pyramidal neurons during movement suggests an involvement in determining the force of contraction. A number of interactions occur between neighbouring columns, including mutual inhibition — a variation of surround inhibition — which may have the effect of sharpening and focussing the motor output,an advantage in fine motor control.

9.4.5   Relationship with sensory input

The nature of interaction between tactile input and motor output has been described already in Chapter 5. One can observe an almost parallel evolutionary development between the pyramidal motor system and the sensory system which conveys this tactile information, the dorsal column medial lemniscal system. As described in section 9.2.1, proprioceptive information from the muscle spindle is also used in finely controlled movements.

9.5   CEREBELLUM

The cerebellum in humans is a fist-sized piece of neural tissue located in the angle between the under surface of the occipital lobes of the cerebral cortex and the upper surfaces of the pons and medulla. Three pairs of legs, or peduncles, connect it to the brainstem. Like the cerebral cortex, the cerebellum is thrown into a series of folds or folia, thereby increasing the available functional surface area. Anatomically a number of regions can be identified. On either side of the central vermis lies a lateral lobe which may be subdivided into anterior and posterior lobes. At the posterior aspect is the flocculonodular lobe. The folia of the cerebellum are much more regularly arranged than the gyri of the cerebral cortex, each running in the same transverse plane. Similarly the neuronal machinery is apparently much more regularly arranged than in the cerebral cortex, which has led to superficial comparisons of cerebellar circuitry with that in computers. Buried inside the cerebellum lies a group of deep cerebellar nuclei which receive much of the cerebellar output and relay it elsewhere.

9.5.1   Cerebellar functions

The cerebellum seems to be involved in several aspects of motor function (Table 9.1). These include balance, adjustment of sensitivity of various reflexes, and the

Table 9.1 — Summary of effects of lesions of cerebellum

Decomposition of movement
Staggering gait
Slurred speech
Loss of upright posture
Tremor

learning, co-ordination and execution of skilled movements. Lesions, depending on the exact site, cause disturbances of balance, tremor, and a loss of ability to produce smooth, skilled, co-ordinated movements.

9.5.2 Inputs and outputs

The cerebellum receives inputs from collaterals or branches of both large and small pyramidal cells of wide areas of the cerebral cortex, including the primary motor sensory cortex. These come via relay nuclei in the pons and form the mossy fibre input which excites both the cerebellar cortex and the deep cerebellar nuclei. Inputs also arrive via the mossy fibre system from major sensory systems, including proprioceptive, visual, auditory and tactile afferents. Thus it is a point of convergence of motor output (desired movement?) and sensory input (information on actual movement?).

A major destination of cerebellar output is, via the deep cerebellar nuclei and thalamus, back to the motor cortex. One of its possible functions is to generate an error signal based on a comparison of the intended with the actual movement, which is then relayed back to the motor cortex.

9.5.3 Cerebellar structure — function relationships

The major output neuron in the cerebellar cortex is the Purkinje cell, whose axons project onto the deep cerebellar nuclei and parts of the vestibular nuclei where the effects, interestingly, are entirely inhibitory. Thus the cerebellar cortex seems to act by an inhibitory interaction with the excitatory effects on the deep cerebellar nuclei from motor cortex and sensory inputs. It is this inhibitory fine-tuning of motor commands which seems to be the critical aspect of cerebellar function.

9.5.4 Purkinje neurons

The Purkinje cells are found in the middle layer of the three layers of the cerebellar cortex and, in contrast to neurons in most parts of the nervous system, are lined up in rows. These rows run along the length of the folium. Their dendritic trees are almost planar, perpendicular to the long axis of the folium and extending up into the most superficial layer of cortex.

9.5.5 Mossy fibres

The mossy fibre input synapses with a neuron known as a granule cell in the deepest of the three cortical layers (Fig. 9.8). The axon from this ascends to the most superficial layer of cortex, where it bifurcates to form the so-called parallel fibres

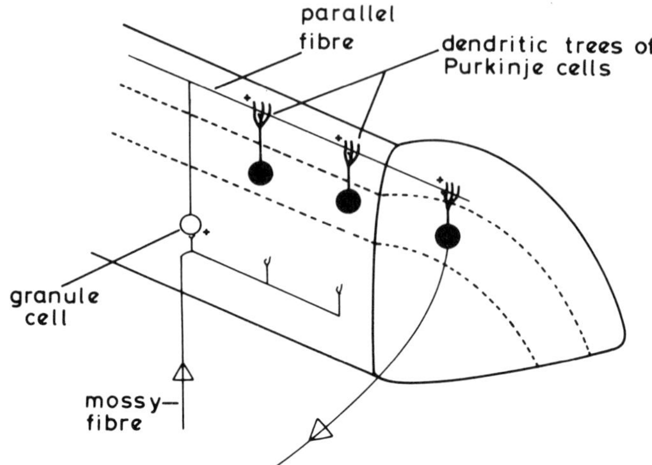

Fig. 9.8 — Parallel fibre system of cerebellum cutting through a line of Purkinje cells.

which run for several millimetres in either direction parallel with the long axis of the folium and thus perpendicular to the plane of the dendritic tree of the Purkinje cell.

A single parallel fibre cuts through the dendritic trees of a row of around 50 Purkinje cells, while each Purkinje cell dendritic tree is innervated by about 250 000 parallel fibres.

Collaterals, derived from the parallel fibres, also innervate a number of types of intracortical inhibitory neuron (Fig. 9.9). Amongst these are basket cells, which

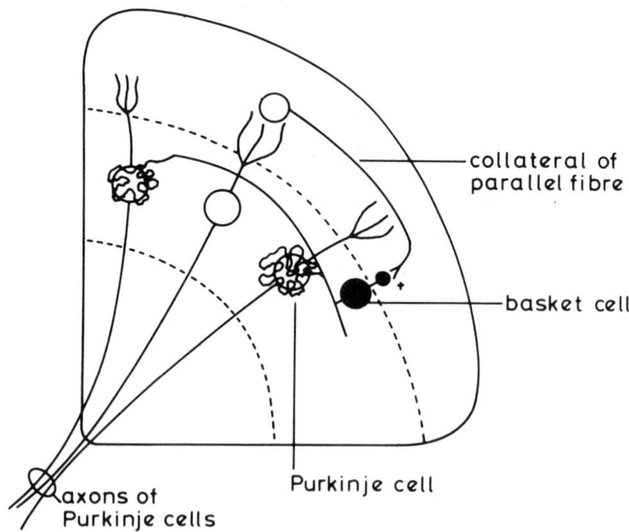

Fig. 9.9 — Basket cells in cerebellum exerting surround inhibition on neighbouring lines of
Purkinje cells.

powerfully inhibit adjacent rows of Purkinje cells, — yet another instance of surround inhibition — and Golgi cells, which produce feedback inhibition of the granule cell (Fig. 9.10).

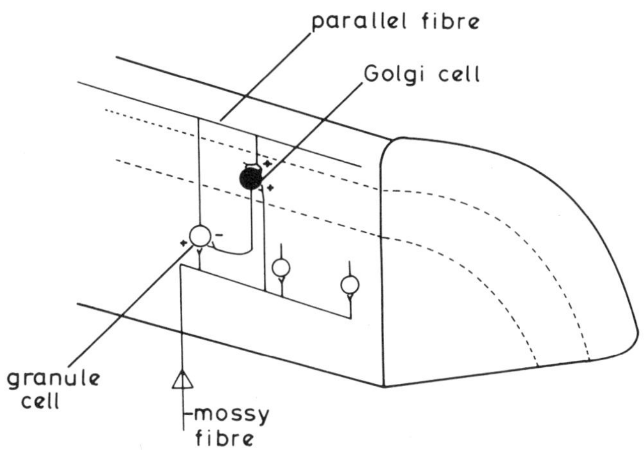

Fig. 9.10 — Feedback inhibition on cerebellar input.

Activation of the mossy fibre system will therefore activate particular rows of Purkinje cells, depending on the interaction of the movement command from motor cortex and subsequent sensory feedback. The result is the dispatch of specific patterns of action potentials to the deep cerebellar nuclei.

9.5.6 Climbing fibres

The cerebellum possesses a second type of input. This is via the climbing fibre system. It arises from small pyramidal cells of the motor cortex and reaches the cerebellum after synapsing in the inferior olivary nucleus. The distribution of climbing fibres and the effects on Purkinje cell discharge are quite different to those of mossy fibres. Each climbing fibre supplies fewer than 10 Purkinje cells,each of which receives innervation from only one climbing fibre. The nature of the synaptic contact is interesting. The terminals climb all over the soma and wrap around the entire dendritic tree, making extensive contact (Fig. 9.11).

Climbing fibre activity seems to be essential for the maintenance and development of the Purkinje cell dendritic tree. Activation of the Purkinje cell by its climbing fibre results in complex potential changes of the Purkinje cell. Experiments have shown that severance of the climbing fibre input results in a rapid involution of the dendritic tree. Further, it has been shown in primates that climbing fibre activity is maximal during the acquisition of a new motor skill. If one accepts that a memory trace or engram results from the establishment of preferred pathways (see Chapter 12) and that development of the dendritic tree is related to this, then these observations would tend to suggest that one of the functions of the climbing fibre

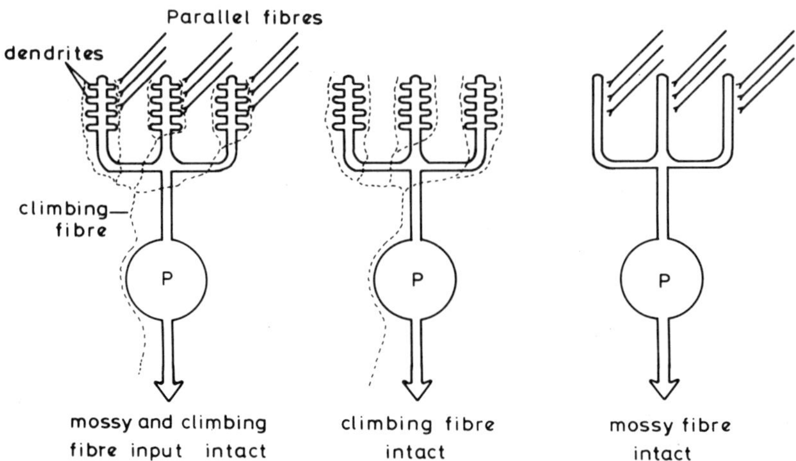

Fig. 9.11 — Effects of severance of climbing or mossy fibre input on dendritic tree of Purkinje cell. (After Hamori, J. (1972) Developmental morphology of dendritic postsynaptic specializations. In: *Recent development of neurobiology in Hungary, Vol. 4*. K. Lissak (Ed.), Akademia Kiado, Budapest).

system might be to instruct sequences of Purkinje cells to learn particular motor skills.

9.6   BASAL GANGLIA

Deep within the cerebral hemispheres, close to the thalamus, lies a group of nuclear masses collectively known as the basal ganglia, which are an important element in the extrapyramidal system. These comprise the globus pallidus and the corpus striatum, the latter being divisible into caudate nucleus and putamen (Fig. 9.12). It is clear from clinical studies that these nuclei are important in motor control, but precise descriptions of how they are involved is even more difficult than for the cerebellum. Functionally related to the basal ganglia are a number of other nuclei: the ventrolateral nucleus of the thalamus, the subthalamic nucleus, the red nucleus and the substantia nigra (Fig. 9.13).

9.6.1   Inputs and outputs

Most of the input to the basal ganglia is to the caudate nucleus and most of the output from the globus pallidus. Between these two points there is considerable complexity of interconnections between the various nuclei and their functionally connected neighbours.

The basal ganglia receive major inputs from the limbic system and cerebral cortex, while major outputs go to the reticular formation and, via the ventrolateral nucleus of the thalamus, back to the motor cortex. There are thus well-developed connections between the basal ganglia and motor cortex.

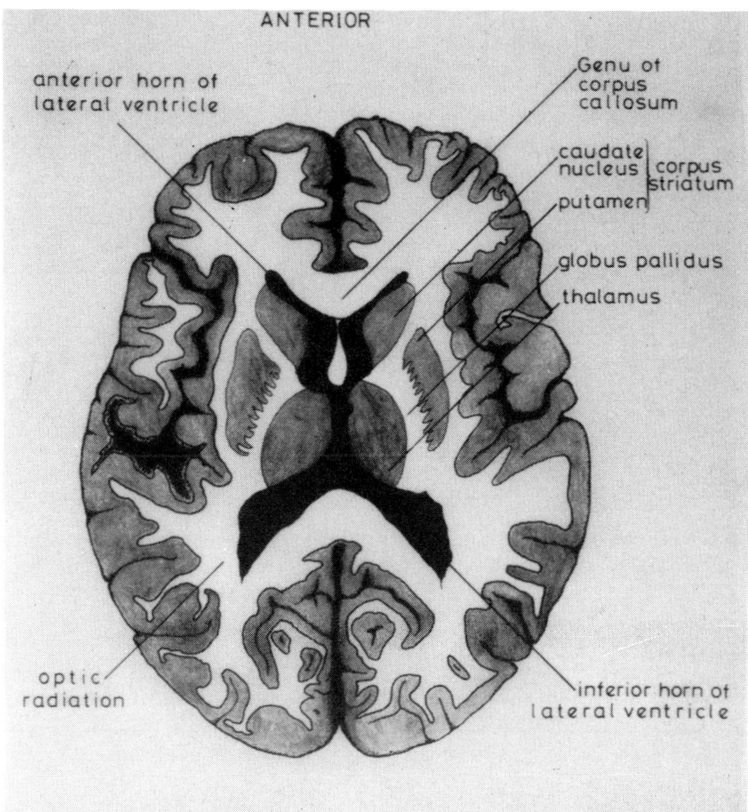

Fig. 9.12 — Horizontal section through cerebral hemispheres to show deeply buried basal ganglia. Figure drawn from a specimen stained with copper sulphate to show up the nuclear masses.

9.7 EFFECTS OF LESIONS

Again, one way to attempt to assess the contribution of the basal ganglia to motor control is to examine the consequences of their malfunction (Table 9.2). A number of human motor disorders have been ascribed to basal ganglia malfunction. Depending on the site of the lesion within the basal ganglia, the symptoms range from hyperkinetic–dystonic afflictions resulting in the production of excessive, involuntary movements, to hypokinetic–rigid syndromes in which there is an attenuation of the ability to produce voluntary movement.

9.7.1 Parkinson's disease

One of the most documented disorders is Parkinson's disease, named after the nineteenth century physician who fully described it. The major symptoms include hypokinesia — a slowness in the initiation and execution of movements that may involve the facial muscles, resulting in a loss of facial expression. The other major symptoms are an increase in muscle stiffness (rigidity) and a resting tremor.

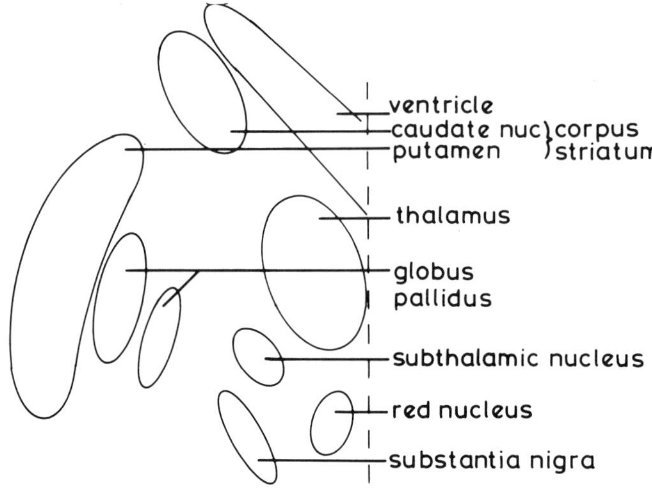

Fig. 9.13 — Relationship of basal ganglia with some other functionally related nuclei.

Table 9.2 — Summary of effects of lesions of basal ganglia

Either	*Or*
Hypokinetic–rigid syndromes	Hyperkinetic–dystonic syndromes
e.g. Parkinson's disease	e.g. Athetosis
	Huntington's chorea
	Hemiballism

Causative agents range from idiopathic (unknown), through viral (e.g. following the pandemics of encephalitis lethargica) to drug-induced (e.g. due to the neuroleptic drugs used in the treatment of schizophrenia).

The common strand seems to be a loss of dopamine production in the CNS. The most important site of damage seems to be in the substantia nigra in which dopamine-producing cells are either destroyed or prevented from synthesizing dopamine. The axons of these neurons normally project to, amongst other places, the corpus striatum, where they release dopamine, and, probably via an interneuron, inhibit the neurons there. Symptomatic treatment is by administration of the dopamine precursor L-DOPA (since dopamine itself does not easily cross the blood–brain barrier), which is taken up by the remaining substantia nigra neurons and converted to dopamine. Such treatment can provide symptomatic relief for many years. However the disease process is progressive and drug therapy may eventually become ineffective.

Recent work has opened up the possibility of implanting dopamine-producing neurons originating either from the donor's own body (e.g. adrenal medulla) or from foetal material. The tremor of Parkinson's disease is of interest in that it consists of reciprocal excitations of flexors and extensors with a frequency of around 5 Hz. It seems to arise as a result of the disrupted output of the basal ganglia to the ventrolateral thalamus where the neurons have a natural rhythm of 5 Hz. In the Parkinson patient neuronal discharges here tend to become synchronized, with a 5 hz period. Lesions of the ventrolateral nucleus of the thalamus have been used in the surgical treatment of Parkinsonian tremor.

9.7.2   Huntington's chorea

In addition to tremor, involuntary movements of many types can arise from lesions in various parts of the basal ganglia. These range from choreiform (dancing) through athetosis (writhing, snake-like) to ballistic (rapid limb extensions). Huntington's chorea, a genetically determined disorder accompanied by rapid intellectual deterioration, was fully described by the American physician George Huntington. Its existence in the United States can be traced back to a single carrier family of immigrants in the seventeenth century. Until recently it was not possible to determine whether a particular individual carried the gene until symptoms appeared, usually after the age of 30, by which time it may have been passed on to children. Recently, a gene probe has become available to test for the presence of the gene.

9.7.3   Hemiballism

Hemiballism, in which there are rapid, violent movements of the limbs of one side of the body, is associated with lesions in the subthalamic nucleus.

9.7.4   Gilles de la Tourette syndrome

One important aspect of motor function which is often overlooked is the control of the muscles involved in speech. One rare disorder which may involve basal ganglia lesions is Gilles de la Tourette syndrome. In this there are frequent involuntary vocalizations, which range from inarticulate cries and grunts to words and phrases, sometimes of an obscene nature.

9.7.5   Functions of basal ganglia

Lesions of the basal ganglia, therefore, seem to cause problems with the planning and initiation of movements, resulting in either the production of unwanted movements, or the inability to initiate a movement voluntarily. It may be from here, therefore, that the first, crude program for a motor act is issued, particularly the ballistic elements. It is tempting to speculate that the explosive expletives uttered in la Tourettes syndrome are the vocal equivalent of a ballistic movement of a limb.

Additionally, a number of important reflexes associated with movement are controlled here, such as staggering when the upright posture is disturbed by a horizontal force. It could be that a motor act cannot take place until all the various reflexes associated with its production have been assembled.

In summary, then, motor activity is a complex mix of movement and postural changes, requiring contributions from many parts of the nervous system, and arranged in a hierarchical though interactive manner. Although the decision to

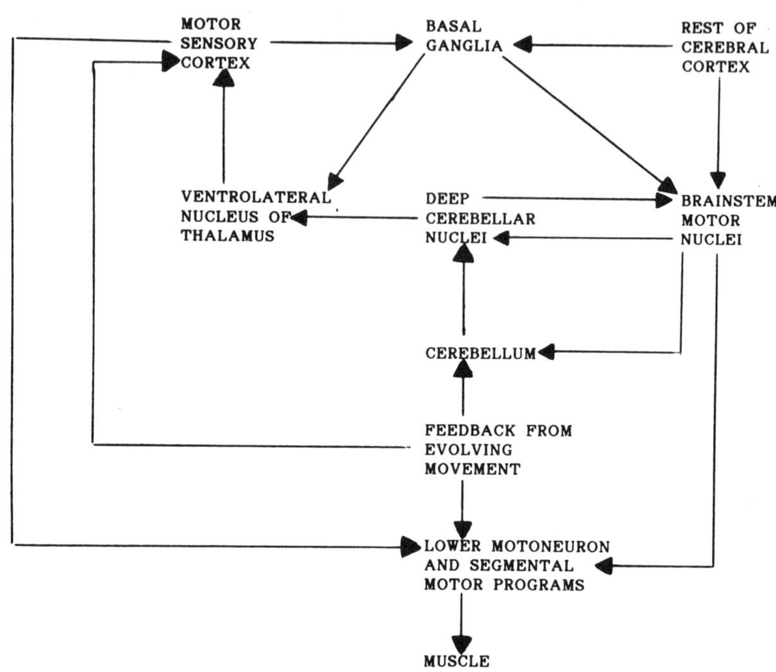

Fig. 9.14 — Summary of interactions between motor control centres.

initiate a movement or series of movements is made centrally, the importance of
sensory feedback in the subsequent execution and control of that act is stressed in
this, and in other chapters. A flow diagram summarizing the interactions between the
various motor control centres is shown in Fig. 9.14.

10

Autonomic nervous system

So far we have considered only the somatic nervous system and how it controls skeletal muscle in the production of voluntary behaviour. The autonomic nervous system (ANS), on the other hand, is responsible for the regulation of visceral function. The gastro-intestinal and urinary tracts, cardiovascular and respiratory systems all come under its control. It may, depending on circumstances, provide for homeostatic or adaptive responses of the viscera.

Homeostasis is the stabilization of internal conditions within narrow ranges by the use of compensatory regulatory mechanisms. Adaptive responses are those triggered by a change in conditions. An example of the former would be the maintenance of heart rate within a particular range during relative indolence, while the onset of vigorous activity would require an adaptive increase in heart rate.

We have within us an immensely complex internal environment. The French physiologist Claude Bernard was the first, around the middle of the last century, to consider the concept of homeostasis: how we maintain the constancy of this internal environment in terms of blood pressure, ionic distributions and metabolic requirements.

Although normally this is achieved automatically, without our conscious intervention, we can train ourselves to exert some conscious control. A good example is reduction of blood pressure. Superimposing conscious control of blood pressure is an important element in meditation techniques, allowing its reduction in people in whom it is dangerously high.

In many cases the simple physiological response requires some form of behavioural correlate (Fig. 10.1). Homeostatic regulation, for example, may need to be supported by the ingestive behaviours of eating and drinking. We must periodically replace the water we continuously lose through respiration, urination and perspiration and the food we metabolize to provide us with the energy to think and move around. The ANS is involved in generating the drives of thirst and hunger which stimulate the behavioural responses of drinking and eating, and is also involved in determining the reward element in satisfying that need. Research in this field gives important insights into eating and drinking disorders and into aspects of drug addiction.

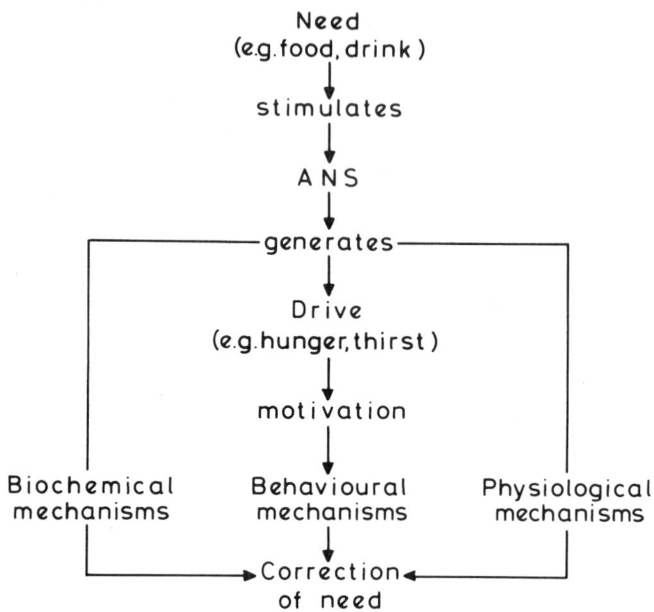

Fig. 10.1 — Involvement of ANS in homeostasis.

Adaptive responses are triggered by the ANS following changes in conditions. One example of such changed conditions has already been given at the sudden onset of exercise. To accomplish increased activity our muscles need more oxygen. This is supplied by ANS-induced increases in the force and rate at which the heart beats. This physiological response would be useless if more glucose was not made available for oxidation. Thus biochemical changes also form part of the integrated response.

Let us suppose now that the need for an adaptive response was triggered not by our voluntary desire to engage in exercise but was imposed upon us by some external threat. The ANS is also concerned with the integration of these behaviours. A good example is the 'fight or flight response' triggered when danger threatens. These biochemical and physiological responses are accompanied by emotional responses of aggression, fear and anxiety. So well-integrated are the various levels of response that it is exceedingly difficult to unpick cause and effect between the physiological and emotional responses. For example, do the hairs standing up on the backs of our necks and our thumping heart cause, or are they generated by, the accompanying fear? Controversy still surrounds peripheral versus central theories of psychophysiological responses.

Elements of the ANS are represented within the CNS in spinal cord, brainstem and forebrain. There are particularly important control areas in the hypothalamus, thalamus and parts of the limbic system such as the amygdala. The hypothalamus is a particularly important ANS integrating centre and has close links with the endocrine system. Let us first consider the distribution and tissue-specific actions of the ANS.

10.1 DISTRIBUTION

The ANS may be divided functionally and anatomically into sympathetic and parasympathetic components. Peripheral distribution of the sympathetic part comes from thoracic and lumbar segments of the spinal cord while the parasympathetic arises from the brainstem, hypothalamus, and sacral parts of the spinal cord.

After leaving the brainstem or spinal cord, the autonomic nerve enters a ganglion where it may synapse. In the sympathetic system the ganglion is close to the spinal cord with a long postganglionic fibre to the target organ, whereas in the parasympathetic system the preganglionic fibre is long with the ganglion lying close to, or buried within, the target organ.

Most target organs are innervated by both ANS components on which they exert opposite effects (Table 10.1). Thus the net activity of any dually innervated tissues or

Table 10.1 — Summary of actions of ANS on specific organs and tissues

Locus of action	Sympathetic	Parasympathetic
Pupil of eye	Dilation	Constriction
Cardiac muscle	Increase rate and force of contraction	Decrease rate and force of contraction
Skin — sweat glands	Stimulates sweat production	No parasympathetic supply
Skin — blood vessels	Constriction	
Skin — hairs	Erection by stimulation of pilomotor muscle	
Adrenal gland	Secretion	No parasympathetic supply
Bronchi of lungs	Dilation	Constriction
Gut	Decreased activity	Increased activity

organs will be achieved by the balance between the two inputs, which can be changed according to circumstances. Which input predominates under resting conditions depends on the tissue innervated.

Some structures, however, have only a single innervation. A good example of this is the medullary part of the adrenal gland. This is a modified and much enlarged sympathetic ganglion which has taken on an endocrine function and releases the hormones adrenalin and noradrenalin into the bloodstream during periods of activity. It is highly active, for example, as part of the 'fight or flight' response.

10.2 PHARMACOLOGY

The ability to pharmacologically manipulate ANS activity endows enormous human benefit and has been intensively studied. For example, the benefits of controlling high blood pressure (strokes) or the amount and timing of acid and digestive enzyme release by the gut (ulcers) are clearly enormous. The major transmitter used by the preganglionic fibre is acetylcholine. On release this combines with nicotinic receptors on the postganglionic neuron. The major transmitter utilized by the postganglionic sympathetic fibres is noradrenalin while acetylcholine is again used by the parasympathetic fibre. Here, however, it combines with the muscarinic type of cholinergic receptor on the target organ. Let us now consider in more detail some of the integrating actions of the ANS.

10.3 HOMEOSTASIS

In homeostasis a number of common elements are required, regardless of the variable which is being controlled (Fig. 10.2). Let us illustrate these with reference to

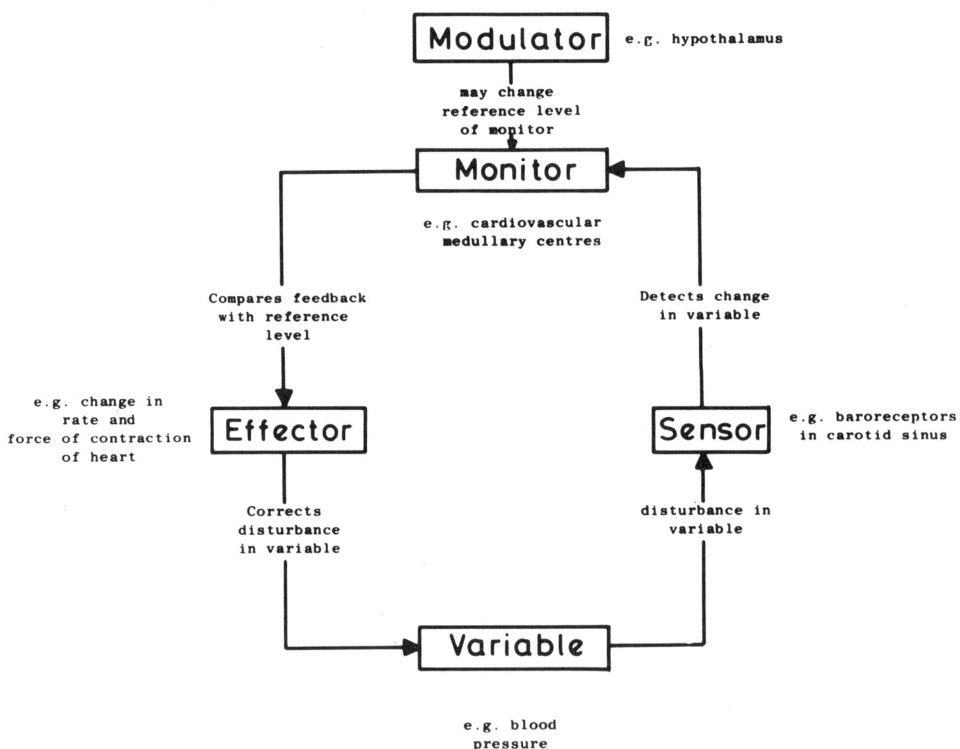

Fig. 10.2 — Components of a homeostatic control system.

the regulation of blood pressure. Some type of sensor is necessary to detect any change in the variable being controlled. In the case of blood pressure these are the baroreceptors, a kind of stretch receptor found in the walls of the aortic arch and at the bifurcation of the internal and external carotid arteries. Any increase in blood pressure will distend these vessels stretching the baroreceptors. Somewhere in the CNS is a monitor which compares the feedback from the receptors with the reference level. In the case of blood pressure these are the cardiovascular control centres in the brainstem. On the basis of this comparison the monitor dispatches instructions to an effector to reverse the change in the variable. For blood pressure the effector is the heart. Increases in blood pressure thus lead to a decrease in heart rate, while blood pressure decreases result in an increase in heart rate.

If blood pressure is so rigorously regulated, how then can we adapt to changed circumstances, such as the need for fight or flight? The answer is that the cardiovascular monitoring system in the brainstem can be overidden or its reference level changed. The hypothalamus is important in such changes and is directly involved in other aspects of homeostatic control and in some adaptive responses.

10.4 HYPOTHALAMUS

The hypothalamus is a phylogenetically ancient part of the brain found in the base of the third ventricle, limited anteriorly by the optic chiasm, posteriorly by the mamillary bodies. It is divisible into medial and lateral portions, the medial part of which is organized into a number of neuronal concentrations called nuclei (Fig. 10.3). Its major activities seem to be the organization of homeostatic and adaptive

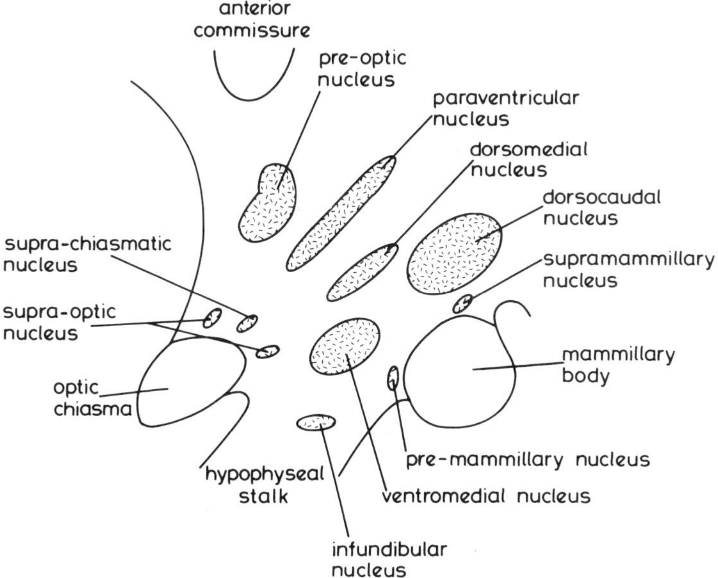

Fig. 10.3 — Nuclei of hypothalamus.

responses and their integration with somatic mechanisms. It regulates body temperature, fluid balance, metabolic substrates, sex, and sleep (not necessarily in that order).

For a number of functions the nuclei are paired on a reciprocally antagonistic basis. One of the pair would provoke reduction of a raised variable such as body temperature, while the other would cause its elevation from a reduced value.

It exerts functional control over the pituitary gland, which hangs by a narrow stalk from its base. The anterior part of the pituitary is controlled by means of the hypophyseal portal system: a capillary network running through the pituitary stalk into the anterior pituitary. The hypothalamic neurons secrete releasing factors into this local network, which are carried to the anterior pituitary and affect the cells there.

The posterior pituitary contains the terminals of axons of hypothalamic neurons. These end close to the capillary bed of the posterior pituitary. Their secretions are released into it and pass into the general circulation. Since the pituitary gland in turn controls many of the body's endocrine glands, the hypothalamus exerts considerable influence over endocrine function. The hypothalamus in turn is controlled from higher centres, in particular from the limbic system.

10.4.1 Body temperature
After extensive hypothalamic lesions a mammal becomes unable to maintain a constant body temperature in the face of variations in environmental temperature changes. Mechanisms to reduce an elevated body temperature are initiated by a nucleus in the anterior hypothalamus. These include initiation of sweating, vasodilation of skin blood vessels and reductions in metabolic rate. A nucleus in the posterior hypothalamus initiates heat conservation responses provoked by a fall in body temperature. These include vasoconstriction of skin blood vessels, an increase in metabolic rate, piloerection (elevation of the fine skin hairs, which results in goose bumps) and shivering — rapid involuntary muscle activity which releases heat, helping to raise body temperature. Feeling uncomfortably hot or cold also gives us the motivation to make appropriate behavioural responses.

10.4.2 Thirst, ionic and fluid balance
Destructive lesions in specific parts of the hypothalamus result in diabetes insipidus, a condition characterized by an inability to control the amount of water lost in the urine, which may amount to 25 l per day. It is accompanied by a raging thirst which stimulates the sufferer to make good the water lost by copious drinking. The hypothalamus is responsible for control of both water loss and water gain.

We need to maintain the fluid content of our various bodily compartments for a variety of reasons, one reason being regulation of the concentrations of the various intracellular and extracellular ions on which life depends. Since we are continuously losing water through perspiration and in urine and expired gases, it must be replaced. Mechanisms are triggered to minimize water loss and stimulate water gain. Decreases in plasma volume or increases in ionic concentrations in bodily fluids indicate loss of water. One of the hypothalamic nuclei detects these changes and triggers two important responses. The neurosecretion antidiuretic hormone (ADH), synthezised in neurons in the hypothalamus, is released into the bloodstream from their axons, which terminate in the posterior pituitary gland. Its major effect is on the kidney tubules, where it increases the reabsorption of water. The hypothalamic nucleus which triggers ADH release also acts on another hypothalamic nucleus to generate the sensation of thirst. Thus the need for fluid intake generates a drive in the form of thirst which motivates us to drink.

Few of us, when thirsty, lust for a glass of water. Some of the fluids we drink somewhat disturb this beautifully integrated physiological and behavioural balancing mechanism. Caffeine, for example, the CNS stimulant present in coffee, inhibits the release of ADH. This explains the urinary urgency which overcomes us shortly after drinking a cup. Alcohol is another potent ADH release inhibitor. Many of us are familiar with the 'pint of beer' syndrome. If one's thirst is quenched with a refreshing pint of beer, the alcohol in it suppresses ADH production, leading to further water loss. This makes the imbiber thirsty to drink another pint. The cycle may be repeated according to the enthusiasm of the participant.

10.4.3 Food intake

The energy stores we expend during our various mechanical and electrochemical activities must be periodically renewed by the intake of food. Lesions in the ventromedial nucleus of the hypothalamus have been associated with hyperphagia and subsequent obesity, while damage in the lateral hypothalamus result in aphagia and weight loss. These areas became known as satiety and feeding centres respectively. In some species stimulation of the lateral hypothalamus will evoke not only feeding, but food-finding behaviour, such as stalking.

The balance between the two seems to be with the feeding centre. Lesions here result in aphagia even in an animal previously made hyperphagic by damage to the ventromedial nucleus. There is controversy as to whether the symptoms produced result from destruction of hypothalamic neurons or whether such treatment severs nerve fibres from elsewhere which are important in feeding behaviour and are simply passing through or to these parts of the hypothalamus. Many different receptors signal our food requirements. These include stretch receptors monitoring gastric distension, taste and mechanoreceptors in the mouth, and glucoreceptors in the liver and hypothalamus.

10.4.4 Fight or flight reaction

When danger threatens biochemical, physiological, psychological and behavioural responses occur collectively known as the fight or flight reaction. Observations of the physiological effects are consistent with a sudden massive shift in autonomic activity in favour of sympathetic output. The physiological changes are widespread. The rate and force of cardiac contraction are increased, which provides more blood for the active muscles. Vasoconstriction of skin blood vessels is accompanied by vasodilation of vessels to skeletal muscle and the heart. This diverts more of the extra blood to active muscle and reduces blood loss from any superficial wound sustained. The pupil of the eye becomes dilated and the fine hairs on the body and neck stand on end. The adrenal medulla is stimulated to release adrenalin and noradrenalin, which increase and sustain these effects.

Adrenalin also has major biochemical effects which result in increased availability of glucose for oxidation. It does this by activating one of the enzymes in the glycolytic chain.

The appropriate emotional responses may also be integrated in the hypothalamus. Electrical stimulation of particular parts of the hypothalamus in cats evokes expressions of rage — hissing, spitting and attack — whereas stimulation of other parts causes the animal to take flight.

The hypothalamus in turn is under modulatory control from the limbic system, for example from the amygdala. Chronic implantation of electrodes in the amygdala for the control of emotional excitability in violent offenders is a controversial area of psychosurgical mind control.

10.5 ANTICIPATORY RESPONSES

With increasing encephalization the initial trigger for the production of a homeostatic or adaptive response may not necessarily come from receptors signalling a deficit of some vital substance, such as glucose or oxygen, but may be consciously anticipated. For example, when we are about to engage in some energetic activity, our rise in heart rate precedes the onset of the activity simply due to its contemplation. Often we eat not because we . e hungry, but because habit dictates that we normally eat at that time of day. As far as food intake is concerned we often rely on long-term strategies rather than pure physiological feedback. An experiment to illustrate this involved feeding a group of rats three meals per day at fixed times. After this pattern had been established the middle meal was discontinued. In the short-term the rats compensated by eating more in what had been the third meal, but eventually started to eat more for the first meal instead. The interpretation of this is that while short-term regulation may be on an automatic basis, long-term regulation involves anticipation of the likelihood of the next mealtime.

11

Cortical localization, lateralization and speech

Observations made during the last century on CNS lesions which cause human speech loss triggered the development of ideas about cortical localization of function. Speech disorders were found to be strongly associated with damage to particular parts of the frontal cortex. Many of the primary sensory receiving and motor output areas of the cortex have subsequently been accurately delineated.

11.1 LOCALIZATION OF FUNCTION

The primary cortical receiving areas for many of the sensory modalities are apparently well localized, even down to specific gyri within a particular lobe. For example, visual afferents project to the striate cortex of the occipital lobe, auditory afferents to the superior gyrus of the temporal lobe and tactile afferents to the postcentral gyrus of the parietal lobe. Similarly, on the output side, the primary motor sensory cortex is in the precentral gyrus of the frontal lobe. Even in these apparently straightforward examples, however, things are not all they may seem. For example, the primary motor sensory cortex receives much tactile input, while considerable motor output originates from the primary somatosensory cortex. Since the brain works in a highly integrated fashion, we might expect functional integration between apparently anatomically localized areas.

When we move to higher mental functions, localization becomes even more difficult to demonstrate. It is thought that higher functions are achieved by complex interactions between large numbers of the more basic, better localized components, together with subcortical structures. It follows that precise anatomical localization at better than lobular level is unlikely to be demonstrated for such activities. We can assign certain aspects of personality and social behaviour to the frontal lobes, long-term memory to the temporal lobes, and constructional and drawing abilities to the parietal lobes. Any attempt at further resolution would necessitate the description of the precise neural connectivities within and between the various areas contributing to that function.

11.2 LATERALIZATION OF FUNCTION

Lateralization is the extent to which the two hemispheres differ in their ability to perform a particular task. The most obvious example is handedness, although we also have a dominant eye, ear and leg. Most people are right-handed, using this hand for highly skilled motor functions. It follows that the left hemisphere is 'better' at the execution of precise motor skills. This seems to go hand in hand (sorry!) with the left hemispheric specializations in language skills, including speech, while the right is better at non-verbal activities. Let us illustrate the concepts of localization and lateralization of function by a more detailed description of speech.

11.3 SPEECH

Speech is a complex motor activity dependent on a number of control regions in the cerebral cortex, basal ganglia, cerebellum and thalamus. The mechanics of speech production will not be discussed here. Suffice it to say that at least three major groups of muscles are involved. A controlled air flow is produced during exhalation by elastic recoil of the thorax and action of the expiratory muscles. This hot air is turned into sound by the muscles controlling the rate at which the vocal folds open and close and, after resonating in the various cavities of the head, shaped into words by the articulator muscles controlling the lips, tongue and mandible.

All these muscles have different rates of contraction; for example we can move the tip of the tongue much more rapidly than we can move the mandible. Therefore, in order to arrive at any particular instantaneous vocal configuration will require complex programming of muscular activation.

Speech does not occur in isolation from other physiological requirements. Respiration must also supply us with sufficient oxygen for our current metabolic activities. Thus we cannot dwell unduly in the expiratory part of the cycle, though some individuals would seem to defy this principle. The thoracic cavity is continously subjected to deformation from other movements in which we may be engaged. For instance, we can still speak (admittedly with some difficulty) after a 100-m sprint. Great accuracy of control is required for intelligible speech. Think how such different sounds as 'p' and 'b' employ such similar configurations of the lips. Clearly speech is a very complex motor act, requiring integrated activity in many muscle groups.

11.3.1 Feedback and speech

Accurate control requires accurate sensory feedback. It is not surprising, therefore, that sensory feedback is important for its execution. The three major types of feedback used are tactile, proprioceptive and auditory, where auditory is probably the most important. Tactile input comes from mechanoreceptors in the tongue and oral cavity and proprioceptive from muscle spindles. Most of us have experienced the difficulties of speech upon loss of tactile input when the local anaesthesia used in dental work affects tongue, lips or cheek.

Auditory feedback involves monitoring our own voice as we speak. The importance of accurate auditory feedback can be gauged by experimentally delaying it (Fig.
11.1). In delayed auditory feedback, the subject's voice is replayed to them through

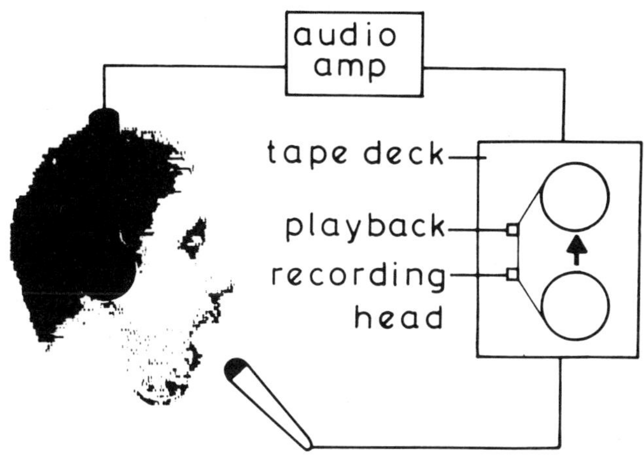

Fig. 11.1 — Delayed auditory feedback.

headphones, but with an additional delay of between 100 and 200 ms introduced.
Changes in the speech are remarkable. It becomes slower, more monotonous,
altered in intensity and pitch, with prolongation of vowels and often production of a
stammer. It seems that there is great difficulty in progressing to the next sound until
the previous one has been monitored and accepted. Faulty auditory monitoring may
contribute to some forms of stammer. The phenomenon of masking has been used
experimentally to prevent auditory feedback. The subject's speech is continuously
analysed for frequency content and white noise of similar frequency played back to
them via headphones. In some instances the stammer is considerably reduced.

11.3.2 Lateralization of speech
Speech is an unusual motor activity in that it requires precise, symmetrical, bilateral
muscle activation. It would be difficult for such precision to be achieved if each half
was under the control of a separate hemisphere.
   In the early nineteenth century Marc Dax observed that speech disorders were
usually associated with left hemispheric damage. The first discovery of which cortical
areas were involved came later in that century when the surgeon Paul Broca
demonstrated at post mortem that two patients who in life suffered from speech
disorders had lesions in the left frontal lobe of the cerebral cortex. Thus there
appears to be functional hemispheric differentiation with a left hemispheric dominance for speech. In the laboratory, such dominance for language functions can be

demonstrated using dichotic listening tests. Different sequences of unconnected words or numbers are played simultaneously, via headphones, to each of the subject's ears. Most people show a better recall of information which enters through the right ear, again indicating a left hemispheric dominance.

Knowledge of hemispheric dominance for a particular function within an individual is of clinical importance for diagnosis and formulation of treatment. For example, prior to neurosurgery, it may be important to determine which hemisphere is dominant for speech. Injection of the short-acting barbiturate sodium amytal into the internal carotid artery of each side in turn will anaesthetize the cerebral hemisphere of that side for 5 to 10 min, leading to a disruption of speech when the dominant hemisphere is affected.

11.3.3   Lateralization and handedness

Cerebral dominance for speech is associated with handedness. Zangwill in the 1960s showed that for both left- and right-handers, aphasia (defects of speech) was more likely to result from left hemispheric damage. However, right hemispheric damage was more likely to result in aphasia in left-handers than right-handers. It makes organizational sense to collect the fine motor skills in the same hemisphere. Why it is usually the left for such motor skills remains to be determined.

This lateralization for speech seems to develop gradually over the first 10 years of life, since it appears that transfer of this function to the non-damaged hemisphere can occur up till this time.

11.3.4   Localization of speech

At least eight cortical regions are involved in the control of speech (Fig. 11.2). These include Broca's area in the anterolateral part of the precentral gyrus, the laryngeal portion of the primary motor sensory cortex, the supplementary motor area on the medial aspect of the hemisphere, the facial part of the primary somatosensory cortex on the lateral part of the postcentral gyrus, Wernicke's area on the posterior part of temporal lobe, the primary auditory cortex on the temporal lobe and extending onto the parietal lobe, the verbal memory cortex on the anterior part of the temporal lobe, and various parts of the parieto-occipital cortex. In addition, of course, the cerebellum and basal ganglia are closely involved. Some idea of the contributions of the various parts can be gained from study of the consequences of lesions.

11.3.5   Effects of lesions

Lesions in the various cortical areas contributing to speech result in a variety of language difficulties known as aphasias.

Damage to the facial region of the primary somatosensory cortex results in afferent motor aphasia. Since there is loss of cortical tactile (and proprioceptive) processing the patient is unable to find the correct position for the tongue, for example, to produce any particular sound.

Efferent motor aphasia results from damage to Broca's area. In this there is no problem in producing single sounds. It is the ability to string sounds together which is compromised. Thus Broca's area seems to be concerned with the serial organization of speech sounds.

Damage to the supplementary motor area results in perseveration, the constant

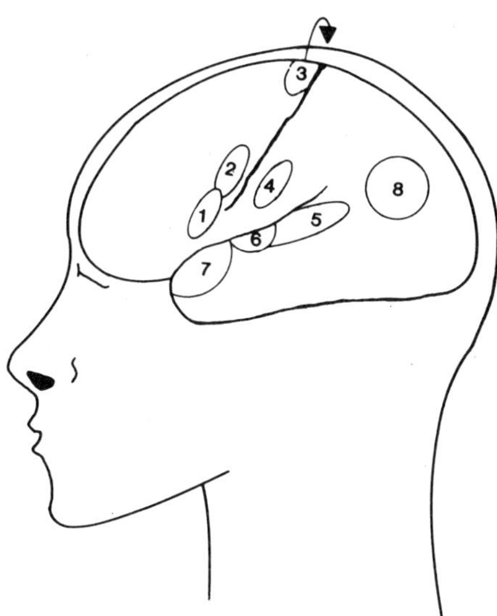

Fig. 11.2 — Major cortical speech areas. 1, Broca's area; 2, motor laryngeal area; 3, supplementary motor area; 4, face/lip/tongue part of primary sensory motor cortex; 5, Wernicke's area; 6, primary auditory cortex; 7, verbal memory cortex; 8, Parieto-occipital cortex.

repetition of the same sound. It may be that this area is responsible for switching the output from one sound to the next.

Damage to Wernicke's area results in acoustic aphasia. In this there is an inability to use auditory feedback to modify speech, associated with a lack of comprehension of spoken text. Ability to analyse complex auditory stimuli, particularly speech, is attenuated. The subject's speech may be fluent, but nonsensical in content — so-called word salads.

Damage to the verbal memory cortex causes the loss of long-term memory of spoken words, while lesions to the parieto-occipital cortex results in semantic reception disorders. In the latter the understanding of what is being said is lost rather than the ability to speak. A flow diagram of the relationships between the various areas is shown in Fig. 11.3.

11.3.6 Wernicke–Geschwind model

This kind of localization and serial organization of various aspects of speech is a very attractive proposition and is well illustrated by the Wernicke–Geschwind model, which proposes that on hearing a word the output from the primary auditory cortex is transmitted to Wernicke's area for matching with verbal memory and comprehension. To speak this word the pattern is passed to Broca's area to serially organize the articulatory form of the sound sequence and from here to the appropriate part of the motor sensory cortex for activation of the muscles of speech.

However, caution must be applied in interpretation from this kind of approach.

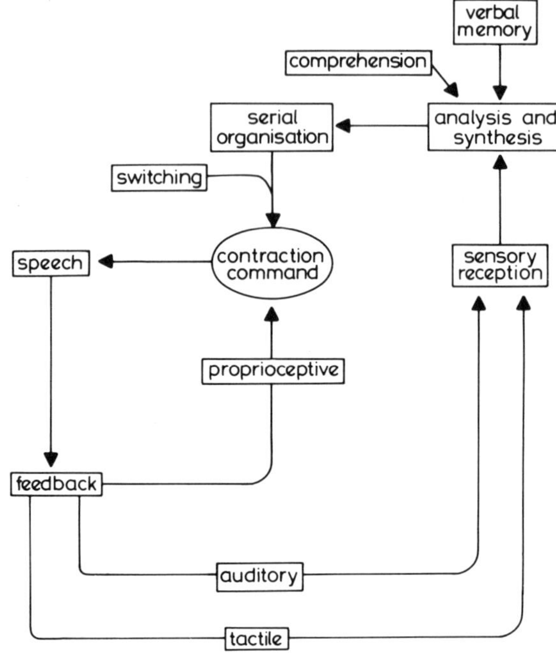

Fig. 11.3 — Relationships between various cortical speech areas.

Powerful arguments can also be assembled for the control of speech on a non-localized, non-serially organized basis. For example, recent experiments using positron emission tomography of spoken responses to visually presented words, which incidentally requires the activity of another cortical speech area, Dejerine's visual language area in the angular gyrus, would seem to indicate that the systems operate in parallel rather than by any single, serially ordered route.

12

Plasticity

The concept of plasticity of the nervous system can be approached in a variety of ways. One aspect is the way in which the nervous system changes in response to experience, its ability to learn and how this abilty is modified by external factors such as nutritional status. Another is the extent to which neural tissue repairs following injury. To some extent the requirements for learning and the capacity for repair are mutually exclusive. The organizational requirements for learning to occur in part contribute to the relatively poor reparative ability. What makes our nervous system such a formidable resource is its ability to learn. Let us therefore start with this aspect of plasticity.

12.1 LEARNING AND MEMORY

Although learning and memory are often thought of in their operation at the conscious level, such as the memory of a face or a piece of music, there are examples of learning in many parts of the nervous system which are not normally associated with conscious memories. Indeed these types of learning have provided much of the evidence for our current views on memory storage at all levels.

We have already encountered one example in the powerful effect of climbing fibre input on the function of Purkinje cells in the cerebellum, possibly associated with the encoding of new motor memories.

Memory formation involves the acquisition of new data from the environment. Implicit in the term is that the memory outlasts the stimulus which caused it. Signal detection and coding have already been discussed in other chapters. What we need to consider here are the neural mechanisms responsible for data storage and how those data may be recalled.

12.1.1 Classical conditioning

Some of the earliest experiments on learning and memory were performed by Pavlov in the early part of the present century. This type of learning became known as classical or Pavlovian conditioning. When we smell or taste food the afferent sensory signals reflexly trigger, via the autonomic nervous system, the flow of saliva, which performs a number of important digestive functions. This normal physiological response to a normal physiological stimulus is an unconditioned response. What Pavlov did was to deliver a neutral stimulus, such as ringing a bell, alongside the physiological stimulus. Eventually salivation, the conditioned response, can be obtained to the neutral stimulus alone. It is thought that the conditioning stimulus becomes incorporated into the memory of the physiological stimulus, eventually forming a sufficiently large part to trigger the physiological response.

12.1.2 Avoidance conditioning

More complex types of behaviour can be evoked by techniques such as avoidance conditioning. For example an animal may be placed in a box to one end of which electric shocks are delivered, preceded by a neutral stimulus such as a flashing light. The shocks can be avoided if the animal performs a certain type of behaviour, such as jumping over a central barrier to the other side of the box. Eventually the animal will execute the learned behaviour in response to the neutral stimulus alone. The next question is at what level in the nervous system can learning occur?

12.1.3 Levels of learning

Evidence suggests that learning can occur at relatively low levels. Horridge, in 1962, showed that the isolated ventral nerve trunk of the cockroach can learn. A cockroach leg, together with an intact nerve supply and its innervating portion of the ventral nerve cord, is isolated from the rest of the body. The leg is placed above a container of an electrically conducting fluid, such that if it remains extended it does not dip down into the fluid. If it becomes flexed, however, it dips into the fluid and receives an electric shock at intervals of a few seconds. Very soon the leg is kept in the extended position. Thus this relatively simple neural circuit has learned to avoid a noxious stimulus.

12.1.4 Stages of human memory

As far as human memory is concerned a number of stages can be recognized. These are known as iconic, short-term and long-term memories. Iconic memory represents the first stage and is most easily demonstrated in the visual system. If a small bright point of light is rotated swiftly in a darkened room a full circle of light can be seen because of the persistence of the visual image. This short lasting memory, of the order of tens of milliseconds, can also be demonstrated in other sensory systems. It is a peripheral phenomenon dependent on the receptor characteristics.

Short-term memory (STM) can be studied by reading aloud sequences of random numbers, letters or monosyllabic words, and asking the subject to recall them. It is found that a sequence of about seven such utterings can be recalled. Some of these STM tasks would seem to be more difficult than others. Thus to remember a sequence of single-digit numbers between 1 and 9 when for any one element the choice is 1 from 9 would seem to be a much easier task than remembering a sequence of letters when there are 26 to choose from or a sequence of monosyllabic words when there are thousands to choose from. However, it seems that a sequence of around seven such utterances can be recalled regardless of type. Thus STM seems to be limited to a particular number of elements rather than the information content of particular elements.

Long-term memory (LTM) contains all the data that have been retained for more than a few moments. It is thought to have enormous capacity and to allow storage for indefinite periods. Since so much information is stored there are sometimes difficulties in retrieval. Many instances when we think we have forgotten something are simply problems of memory retrieval rather than loss of the memory trace. This can be tested experimentally by allowing the subject to observe a collection of different objects. Following their removal the subject is asked to recall the names of as many of the objects as possible. Some of the apparently forgotten items are reshown to the subject who is asked whether the object was previously present. Many more objects can now be named by recognition. Thus the memory is still present, just difficult to recall.

12.1.5 Neural substrates of memory

Can we identify any neural changes associated with memory storage? A number of lines of evidence suggest that STM and LTM are subserved by different neural mechanisms. For example, electroconvulsive therapy (ECT) used in the treatment of depression, interferes with STM, while LTM is attenuated by administration of protein synthesis inhibitors. Observations like these tend to suggest that STM may be reliant on some transient electrochemical phenomenon such as reverberating circuits, while LTM is dependent upon structural changes in the brain. Not all short-term memories are committed to LTM. Parts of the limbic system, in particular the mamillary bodies, hippocampus and cingulate gyrus of the cortex, seem to be critical in the gating and formation of LTM. The complex networks between these various areas is known as the Papez circuit (Fig. 12.1).

12.1.6 Chemical aspects of memory

Early studies of the mechanisms of LTM concentrated on the search for a molecule which was sufficiently complex and variable to store memories, each perhaps embodied in a unique molecular configuration. The discovery of DNA (deoxyribo nucleic acid), and its ability to store and retrieve genetic memories which, by controlling protein synthesis, determine the phenotype of the organism, greatly influenced early workers in the field of molecular aspects of memory.

A type of worm, the planarium, when cut in halves, is capable of regenerating into two whole worms. When one of these worms was trained to curl up in response to an electrical stimulus and then cut in halves, both parts regenerated into worms possessing the curling response to electrical stimulation without further training. In other experiments DNA extracted from trained worms and injected into naive worms made them easier to train. Unfortunately, however, it has not been possible to duplicate these original findings and such ideas on the chemical basis of memory have now been rejected.

Chemical memory of some sort, however, plays an important part in the embryological development of the nervous system, helping to establish particular patterns of connections. One example is the way in which the optic nerve grows and establishes the correct patterns of connectivities with the optic tectum. This can be shown in amphibians, whose regenerative capacities after neural section are much

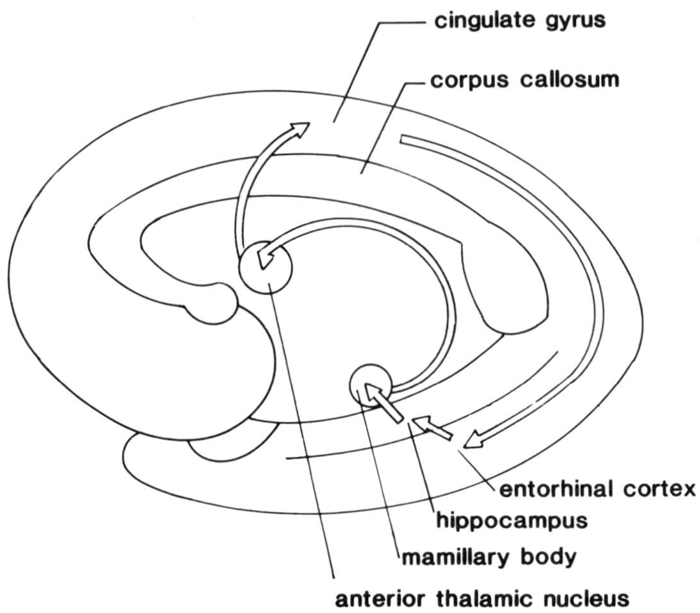

Fig. 12.1 — Papez circuit.

better than those of mammals. The accuracy with which trained frogs strike for flies placed at different positions in the visual field is measured. The optic nerves are then sectioned and the eyes rotated through 180°. Upon regeneration of the nerve the frog's striking accuracy is unimpaired. This is not just a question of the frog learning to deal with a reversed visual field, since if the eyes are rotated without optic nerve section the frog consistently strikes in the wrong direction. Indeed under these circumstances the frog never learns to strike in the correct direction. It must mean that the severed optic fibres re-establish their old connectivities.

12.1.7 Trophic substances
A likely explanation is some sort of chemical coding between points to be connected. The nature of these chemicals is of enormous theoretical significance since adult humans have much poorer regenerative capacities than amphibia. On the other hand, we are much better than the frog at learning reversed visual fields. A human subject wearing inverting lenses very quickly relearns to see the world the right way up.

It would seem that the trophic substances which aid regeneration are present in the human embryo but less obvious in the adult. One such substance, nerve growth factor, a protein originally identified in the brains of chick embryos, has been shown to have powerful effects on the growth of sympathetic neurons.

Substances capable of stimulating correct neural connectivities in other parts of the nervous system would be of considerable human benefit.

12.1.8 Synaptic aspects of memory

Recent evidence would suggest that the major element in the production of LTM is synaptic modification. We have already encountered one example of this in the effect of climbing fibre input on the dendritic tree of the cerebellar Purkinje cell. It should be appreciated that synaptic changes may vary quantitatively, qualitatively and in longevity in different parts of the nervous system, appropriate to the nature of the task that particular neuronal system is carrying out. What they have in common seems to be the establishment of a preferred pathway for information flow, rather like running water gradually etching a river bed along the easiest line of travel.

For example, at the level of the lower motoneuron, various types of potentiation can be observed. A volley of six consecutive action potentials in the axon of a Ia afferent fibre from a muscle spindle, evokes a series of six identical EPSPs in the motoneuron which it monosynaptically innervates. If a similar series of action potentials is induced in a descending pyramidal tract fibre to the same motoneuron, potentiation of the EPSPs occurs, that is the response becomes successively larger to consecutive inputs. The reason for this relatively short-lived potentiation may be connected with the rate of removal of free calcium ions from the nerve terminals after passage of an action potential. This is a rather brief form of memory. Rather longer lasting is the phenomenon of frequency potentiation. An α-motoneuron is stimulated at high frequency for several seconds. The size of the EPSP produced in response to single afferent action potentials is then measured at various times after. It is subsequently found that the EPSP evoked by these single action potentials is greater than normal for a period of more than two hours. Similar experiments performed with hippocampal neurons, which are considered to be intimately involved with conscious memory, show potentiation for more than 10 hours following tetanic stimulation.

12.1.9 Effects of experience on brain development

A somewhat different approach to the study of memory was employed by Rosenzwieg, Bennett and Diamond in the 1970s. They reared groups of rats in a variety of environmental conditions which they termed enriched, standard and impoverished. Those in the enriched environment were given more opportunity for learning by being kept in groups and given lots of objects to play with. Standard rats were kept in groups but without objects to play with. The impoverished were reared singly without objects to play with. When the occipital cortex of each type was examined, interesting differences were observed. There was a correlation between the synaptic development and the quality of the upbringing. Those rats from the enriched environments had synapses which were larger in cross-section with more extensive axodendritic contact. Levels of neurotransmitter biosynthetic enzymes were also higher.

From these sorts of studies growth theories of memory have been developed. It is considered that continuous input to a particular neuron stimulates hypertrophy of the dendritic tree, with proliferation of secondary and tertiary dendritic spines.

Any future traffic coming through that synapse does so more easily and has a more powerful effect on the postsynaptic neuron. It becomes a preferred pathway. Activation of a particular pattern of preferred pathways is probably the basis of the memory trace. Such a memory trace is often referred to as an engram. At the level of conscious memory, localized lesions of the brain do not destroy specific memories but rather common elements of all memories; therefore it is likely that the engram for a particular memory is spread across wide areas of the brain. What is the link between neural bombardment and subsequent synaptic modification? One of the earliest attempts to provide this link was the derepressor hypothesis.

12.1.10　Derepressor hypothesis

At any time large stretches of the DNA code are inactivated by the presence of repressor molecules. Some of these sequences could well be those associated with membrane protein synthesis. It was proposed that particular spatio-temporal patterns of sustained neural bombardment led to a change in the cytoplasmic concentrations of these repressor molecules so that membrane protein and receptor synthesis could proceed. This theory seems to fit quite neatly with the growth theories for the basis of memory. Although a very attractive proposition, there are problems with the derepressor hypothesis in its simple form. One major problem is that many CNS neurons are spontaneously active in the absence of stimulation. This might be expected to lead to modification of all synapses.

Perhaps particular neurotransmitters need to be released or specific pathways activated. For example, the putative neurotransmitter glutamate may be involved in the development of some synapses, since its release has been shown to stimulate dendritic swelling. This action of glutamate is considered more fully in Chapter 13.

Reference to the cerebellar Purkinje cells offers an acceptable modification to the theory in its simple form. It is the input through particular pathways which stimulates synaptic development, in the case of the Purkinje cells, climbing fibre activity seems to have this effect.

12.2　NUTRITIONAL ASPECTS OF BRAIN DEVELOPMENT

The brain is vulnerable to dietary deficiencies at all stages of development. In the adult, long-term vitamin B_1 (thiamine) deficiency leads to Korsakoff's syndrome. Degeneration occurs in the mamillary bodies, and parts of the thalamus. The most obvious symptom is an inability to form long-term memories. It is most often seen in chronic alcoholics, particularly those who also have a poor diet. Iodine deficiency during early brain development leads to cretinism. There is widespread atrophy of the cerebral cortex, accompanied by reductions in axonal and dendritic growth. Vitamin B_6 deficiency leads to myelin sheath degeneration and changes in the metabolism of some central neurotransmitters.

12.3　NEURAL REPAIR

So far we have considered mainly positive aspects of plasticity, ways in which the brain responds and learns. Other aspects, however, concern the reactions following damage to the nervous system. At the neuronal level there is a sequence of short-, medium- and long-term responses. The prospects for functional regeneration depend on many factors, such as whether the lesion is in the CNS or PNS and whether nerve fibres are completely severed or merely crushed. Degenerative changes may be observed not only in the neuron initially damaged, but also in neurons up and downstream from it or in the structures it is innervating. While peripheral neural elements demonstrate considerable capacity for regeneration, that seen within the

CNS is, on the whole, poor. Even within the CNS there is a spectrum of regenerative capacities, with myelinated elements in phylogenetically newer parts of the CNS being amongst the poorest recoverers.

The extent of damage to the CNS caused by an injury is only partly due to direct mechanical damage to the grey and white matter. For example, following acute injury to the spinal cord a number of pathophysiological and inflammatory processes occur. Some of these secondary effects can be reduced by the use of substances such as corticosteroids, leukotrienes and thyrotropin-releasing hormone.

12.3.1   Degenerative changes following nerve section

Augustus Waller in the middle of the last century was the first to describe the changes occurring in a neuron following severance of its axon. The cut ends soon seal and retract from one another. The part of the axon distal to the cut is now isolated from the essential metabolic machinery of the cell soma and undergoes anterograde or Wallerian degeneration, breaking up into ellipsoidal fragments which subsequently undergo lysis. Retrograde changes occur in the section of axon proximal to the cut. While anterograde degeneration following nerve section is inevitable, the severity of retrograde degeneration is very variable, being inversely related to the distance of the lesion site from the cell body. In the worst case there is loss of the coloured Nissl granules, a process known as chromatolysis, and the nucleus takes up an eccentric position. The cell body initially swells then shrinks and is removed by phagocytosis. However, in many instances these retrograde changes may well stop short of neuronal loss and regeneration starts to occur.

12.3.2   Regenerative changes following nerve section

Biochemical changes, such as swelling of the nucleoli, and visible changes, such as a proliferation of rough endoplasmic reticulum, indicating increased RNA and protein synthesis, are amongst the first signs that regeneration is starting.

Axonal regeneration occurs in three phases. These are sprout formation, sprout elongation and maturation. In the first phase tiny branches known as filopodia can be seen sprouting from the cut proximal end of the axon. These fingerlike extensions make constant localized exploratory and contractile movements.

In the PNS, Schwann cells form tubes to guide these branches along. The rate of growth can be quite startling, from 1 to 5 mm per day. Each phase of axonal regrowth is dependent on polymerization of specific soluble proteins to form a fibrillar structure, different proteins being involved in the axonal growth in length and width.

In the CNS axonal sprouting starts but comes to a halt, usually within two weeks. This abortive regeneration is not thought to be due to any intrinsic inability of CNS tissue to regenerate, but rather to a multiplicity of external factors such as the inability of CNS glial cells to form guidance tubes and the absence of cellular-mediated trophic factors or blood-borne factors unable to cross the blood–brain barrier. Damaged neural tissue from the CNS has been shown to regenerate when transplanted into peripheral nervous tissue.

12.3.3   Organizational changes following nerve section

Important organizational responses can also be observed following CNS damage. Work by Wall and Egger in the 1970s illustrates some of these. Having delineated the extent of fore and hindlimb somesthetic representation in the ventrobasal thalamus of the rat, they severed that part of the spinal dorsal columns which was conveying the tactile input from the hindlimb. Thus the ventrobasal thalamic neurons which had been processing hindlimb somesthetic information would no longer have an input from this source. When the mapping procedures were repeated in these animals, it was observed that the hindlimb representation had shrunk considerably. However, what was less expected, was that the forelimb projection had grown. The boundary between the neuronal clusters which had represented fore- and hindlimb respectively had shifted, such that neurons which had previously responded to hindlimb input now fired to tactile stimulation of the forepaw. The conclusion reached was that at the boundary regions between neuronal clusters there is considerable overlap of inputs, which compete for the attentions of the neurons they innervate. Under normal circumstances, one of these inputs, presumably the stronger, predominates. Its synaptic contacts with the post synaptic element proliferate, while the other input is suppressed. In this case the preferred input to these neurons is from the hindlimb. If, however, this stronger input is removed, the other, normally suppressed one can develop.

12.3.4 Mechanisms of organizational responses
Other workers have identified the physical correlates of these observations in terms of synaptic development. One of the processes involved may be collateral sprouting. After the terminal of a degenerating axon has been digested by astrocytic glial cells, adjacent axons form branches known as collaterals, which innervate the vacant postsynaptic site.

So far we have considered the effects of acute, catastrophic neural damage, such as in accidental injury, vascular accidents or surgery. Many disease states, however, result in a more gradual destruction of neural tissue. For example, some neural pathophysiological processes result in a loss of motor innervation to skeletal muscle. As might be expected, this leads to atrophy of the muscles so affected. However, biopsy of the remaining muscle shows a proliferation of the cholinergic receptor. In some cases the receptors proliferate not only in the end-plate region but start to appear all over the sarcolemma of the muscle. Because of this the muscle is much more sensitive to the surviving input. This phenomenon is known as denervation supersensitivity.

Reduction of the amount of neurotransmitter released, caused, for example, by a gradual death of the cell population supplying the transmitter, often leads to a compensatory increase in the numbers of receptors available. For this reason the symptoms of Parkinson's disease do not become apparent until the nigrostriatal neuronal population has been reduced by 90%. Thus in some CNS systems there is enormous capacity to compensate for cellular loss. In Parkinson's disease the typical response due to denervation supersensitivity can be seen. A person in whom the disease had been recently diagnosed would be much more sensitive to exogenously applied dopa than someone without the disease. By the same token, however, pharmacological intervention leading to restoration of the normal neurotransmitter levels will result in a decrease in the numbers of receptor sites available.

13

Neurochemical aspects of mental function

Otto Loewi in the 1920s showed that stimulation of a nerve caused the release of a chemical substance which affected the target organ. In these particular experiments the vagus nerve of an isolated, perfused heart was stimulated, resulting in a slowing of the heart. Addition of the perfusate to a second isolated heart caused this one to slow also. Thus some chemical released into the perfusate around the first heart was responsible for slowing the second heart. Later work showed that this substance was acetylcholine.

13.1   NEUROTRANSMITTERS, NEUROMODULATORS AND CO-TRANSMISSION

Since then it has been shown that a great many types of neuroactive chemicals are employed by the CNS and PNS. In some cases it has been possible with reasonable certainty to identify the transmitter used in a particular pathway by a particular neuronal group. Originally, it was considered that a neuron would release only a single type of neuroactive substance. Recent studies, however, have indicated that in many cases a neuron may release two or more such substances. For example, a neuron may release a 'classic' neurotransmitter substance, having a rapid, short-term postsynaptic effect, together with a neuromodulator, a substance with a slower, qualitatively different postsynaptic effect which may also modify the response to the neurotransmitter. However, we are a long way from understanding the significance and control mechanisms of such co-transmission.

13.1.1   Multiplicity of receptors

The waters are further muddied by the observation that there is more than one receptor type for most transmitters. Thus the same transmitter used by different pathways may well produce different postsynaptic effects because of the presence of alternative receptor types. Receptors are present not only on the dendrites and cell soma but also on axonal terminals. These latter are important in mediating effects such as presynaptic inhibition and as part of feedback mechanisms to monitor and control the amount of neurotransmitter released by the terminal.

13.2 SPECIFIC NEUROACTIVE SUBSTANCES

A comprehensive list of all known and suspected neuroactive agents together with their functions is beyond the scope of this book. A restricted range is described below and summarized in Table 13.1.

Table 13.1 — Table of major neurotransmitters

Putative transmitter/ modulator	Systems/sites used in	Disordered function contributing to
Acetylcholine	Neurotransmuscular junction of skeletal muscle. Basal ganglia Hippocampus	Myasthenia gravis Huntington's chorea Alzheimer's disease
Noradrenalin	Ascending and descending noradrenergic bundles. Assigning value to 'reward' signals of eating, drinking, sleeping, etc.	Depression Sleep disorders
5-Hydroxytryptamine	Brainstem reflexes ? ?	Postanoxic myoclonus Mood disorders Appetite disorders
Dopamine	Nigrostriatal pathway Mesolimbic system?	Parkinson's disease Schizophrenia
γ-Aminobutyric acid	Inhibitory interneurons with short axons in spinal cord, e.g. those producing pre-and postsynaptic inhibition. Output from cerebellum to Deiter's nucleus ?	Spasticity of spinal origin Epilepsy
Glycine	Inhibitory interneurons with short axons in spinal cord mediating postsynaptic inhibition, e.g. Renshaw cells onto motoneurons	
Glutamate	Throughout CNS Post-tetanic potentiation in hippocampus	Memory disorders?
Thyrotropin-releasing hormone	Extrahypothalamic TRH in cerebral cortex, brain stem, spinal cord, etc. Neuromodulator. Co-transmission and interactions with neurotransmitters	Motoneuron disease
Enkephalins	Periaqueducted grey matter. Substantia gelatinosa of spinal cord. Amygdala involved in sensory–emotional integration?	

13.2.1 Acetylcholine

Acetylcholine was the first chemically identified neurotransmitter. Two basic receptor types, nicotinic and muscarinic receptors, mediate its effects. In addition to its importance in autonomic ganglionic transmission and parasympathetic postganglionic transmission, it is used by many pathways in the somatic nervous system. For example it is the transmitter at the neuromuscular junction onto skeletal muscle. The disease myasthenia gravis, which leads to progressive loss of muscle use, is thought to be an auto-immune response against the nicotinic cholinergic receptors here. Each α-motoneuron axon in the spinal cord gives off a collateral branch onto an inhibitory interneuron called a Renshaw cell. Here too the transmitter is acetylcholine. Cholinergic fibres in the corpus striatum of the basal ganglia seem to perform an important balancing effect on neuronal function, opposing the actions of the dopaminergic input from the substantia nigra. Before the advent of L-dopa, an early strategy for the relief of some of the symptoms of Parkinson's disease was the administration of anticholinergic drugs. Cholinergic neurons have been shown to input to the hippocampus. In Alzheimer's disease (presenile dementia) there is an enormous drop in hippocampal acetylcholine content. Whether cholinergic agents or transplantation of cholinergic neurons could provide symptomatic relief remains to be determined.

13.2.2   Noradrenalin

Noradrenalin is released at the terminals of most postganglionic sympathetic fibres. In the somatic nervous system the most obvious pathways utilizing this transmitter are the dorsal and ventral noradrenergic bundles. These arise from neurons in the brainstem and send noradrenalin-releasing fibres to the spinal cord, cerebral cortex, hippocampus, thalamus and hypothalamus. The receptor population for this transmitter are represented by α and β types, each of which is further subdivisible. These pathways are important in the control of blood pressure, mood, and in some way weighting the importance we attach to the reward signals generated by eating and drinking.

13.2.3   5-Hydroxytryptamine (serotonin)

Neuronal cell bodies containing 5-hydroxytryptamine (5-HT) have been found in the brainstem as the origins of both ascending and descending pathways. Electrical stimulation of the ascending component can induce sleep, while one of the effects of stimulating the descending spinal projection is suppression of the flexor reflex in response to a painful stimulus. Thus the descending portion exerts a modulatory influence on spinal pain processing. It is likely that 5-HT is released from a short pathway between the periaqueductal grey matter and nucleus raphe magnus, with some other transmitter being used by the long descending pathway to the spinal cord.

Not only is there a multiplicity of receptors for 5-HT, but also a multiplicity of schemes available for categorizing them. One scheme divides them into types 1, 2 and 3, with type 1 being further subdivisible into a–d subtypes. The suppression of wakefulness and pain would appear to be sensory effects since 5–HT can be shown to be facilitatory on the lower motoneuron, stopping short of overt discharge. These apparent paradoxical effects of reducing sensory sensitivity while making the motor side more excitable might be advantageous during sleep to maintain our motor readiness. Most of us have experienced 'night starts' when, on the edge of sleep, we are galvanized by a trivial stimulus. This may represent a critical switching point in motor excitability between waking and sleeping.

Accidental anoxia for a critical period while under general anaesthesia during surgery can result in a condition known as postanoxic myoclonus. In this, massive reflex muscular responses are triggered by sudden tactile or auditory stimuli. It is thought to be due to a functional loss of 5-HT-mediated brainstem reflexes. The facilitation induced in some neurons by 5-HT seems to be voltage-dependent, having a more powerful effect on the neuron when it is already partially depolarized than when it is at its normal resting potential. In many systems, 5-HT is found to be associated with co-transmitters. For example, in one of the descending pathways controlling motoneuron function, it is released along with the tripeptide thyrotropin-releasing hormone. Malfunctions of 5-HT systems are also thought to occur in disorders of mood and appetite. It is of interest to note that the quantities of 5-HT in the brain are dependent in part on the amount of the precursor amino acid tryptophan which is in our diet. Quite profound and direct relationships might therefore exist between what we eat and our moods and behaviour.

13.2.4 Dopamine

The role of dopamine in the context of movement control and the nigrostriatal pathway have already been discussed. Some patients with Parkinson's disease, when treated with L-dopa, develop psychotic episodes. Some dopaminergic receptor blockers and depletors are effective antipsychotic drugs, in particular in the control of schizophrenia. Theories have been developed linking schizophrenia with overactivity of dopaminergic pathways, possibly those in the mesolimbic system. A number of receptor subtypes have been identified for dopamine, including those designated D1 and D2.

13.2.5 Amino acids

A number of amino acids are thought to act as neurotransmitters. Some of these, such as glutamate and aspartate, have excitatory effects, while others — GABA and glycine — are inhibitory. Glutamate is widely distributed throughout the CNS; so widely, in fact, that some doubt whether it does act as a specific neurotransmitter. Its effect on virtually any neuron it contacts is to produce an intense depolarization. This ubiquitous stimulatory action has long been recognized by food manufacturers who use it as a flavour enchancer. Many pallid products are laced with monosodium glutamate to intensify the stimulation of the taste buds. It has also been shown to have interesting effects on hippocampal neurons which may be related to the mechanisms of STM. A short period of intense stimulation of fibres terminating on these neurons causes release of glutamate from the terminals. The permeability changes induced in the postsynaptic membrane change the transmembrane ionic balance resulting in an osmotic uptake of water. This causes the dendrites and their spines to swell, making the synapse more efficacious for a short period for further information flow. The phenomenon of post-tetanic potentiation, caused in this case by dendritic swelling, may generate a short-term preferential pathway in response to previous stimulation and be the basis of STM in the hippocampus.

The amino acid GABA has a wide distribution throughout the CNS from cerebral cortex to spinal cord. Receptor categories include A and B subtypes. Considerable evidence is available that it acts as an inhibitory neurotransmitter by altering membrane conductance to chloride ions. It is often released by inhibitory interneurons with short axons, and has been shown to mediate both pre- and postsynaptic types of inhibition. Many of the short interneurons in the dorsal horn of spinal grey matter are thought to release GABA. A longer GABAergic pathway runs between the cerebellum and Deiter's nucleus. Loss of the inhibitory effects of GABAergic pathways, may be involved in a variety of conditions, such as epilepsy, Huntington's chorea and spasticity of spinal origin.

The inhibitory amino acid glycine is not as widely distributed as GABA. Feedback inhibition on α motoneurons is blocked by the poison strychnine, a glycine antagonist. Glycine is thought to be released by the inhibitory interneurons Renshaw cells which mediate this recurrent inhibition.

13.2.6　Peptides

A number of larger molecules are thought to act as neurotransmitters/modulators. Amongst these are a number of peptides. One of these is substance P, an 11-residue polypeptide. It was first isolated from brain material in the 1930s but it was not for another 30 years that it was shown to be released from nerve terminals. It is found in particularly high concentrations in the substantia nigra and in the dorsal horn of the spinal cord. In the latter it is located in the terminals of C-fibre afferents, where its release has been proposed to be inhibited by the action of axo-axonic enkephaliner-gic terminals. For these reasons substance P has been proposed as the sensory afferent transmitter for slow pain in the spinal cord. However, almost total depletion of substance P from the cord does not attenuate subsequent noxious inputs. Substance P antagonists block pain of mechanical but not thermal origin. Interest-ingly, high doses of these antagonists produce hindlimb paralysis which might implicate substance P in our behavioural responses to pain rather than its perception. Substance P is also present in the terminals of descending fibres in the spinal cord, where it is co-released with 5-HT. Its actions here are unknown.

Possible involvement of the enkephalins in pain processing has been discussed in Chapter 7.

The tripeptide thyrotropin-releasing hormone, in addition to its endocrinological actions on the pituitary gland, is also widely distributed throughout the nervous system. It is often found in nerve terminals along with, but in separate vesicles from, some other neurotransmitter such as 5-HT, dopamine or noradrenalin. It may have some modulatory effect on neurotransmitter action. For example it has been shown to potentiate the excitatory effects of 5-HT on motoneuron excitability. Its possible use is currently being investigated for the symptomatic treatment of motoneuron disease — a condition characterized by selective motoneuronal degeneration. It may also possess trophic effects on neuronal repair and development.

Glossary of terms

ablation Removal or destruction of part of central nervous system.

absolute refractory period Short period of time following the action potential when the nerve fibre is incapable of generating a second action potential at that point on the membrane.

accommodation reflex Fattening of lens to focus on near objects.

action potential Transient axonal permeability shift in favour of sodium ions. Used for information transmission.

active transport Conduction of ions or molecules across a membrane by a carrier molecule against an electrochemical gradient using metabolic energy.

acuity Ability to see detail.

adaptation Reduction in response of a sensory receptor to continuous or repeated stimulation.

adenosine triphosphate (ATP) Energy storage molecule. Energy is released when broken down to ADP (adenosine diphosphate).

adrenal gland Medullary part is an enlarged sympathetic ganglion modified into an endocrine gland which releases adrenalin and noradrenalin into the blood upon sympathetic stimulation.

afferent Conduction towards some structure.

agonist Chemical substance which has the same effect on a receptor as the natural neurotransmitter.

akinesia Absence of movement.

amino acid Organic molecule of general structure $R–CH–NH_2–COOH$, where R may be a variety of structures. They are the building blocks of peptides and proteins.

amplitude modulation Coding of changes in stimulus characteristics by amplitude of response produced.

amygdala Group of nuclei located in temporal lobe of cerebrum, which are part of limbic system.

anaesthesia Loss of pain plus other modalities.

analgesia Loss of pain without interference with other modalities.

antagonist Chemical substance which has the opposite effect on a receptor to the natural neurotransmitter. Alternatively used to describe a muscle which acts at a joint in a way opposite to that of the reference muscle.

anterior At the front.

anterograde (Wallerian) degeneration Neuronal degeneration distal to axonal section.

antidromic Signal propagation in direction opposite to normal; for instance, impulse conduction in the axon towards the cell body.

aphasia Problems of communication by, or comprehension of, speech or writing as a result of brain damage.

ascending In a rostral direction towards higher parts of the nervous system.

association cortex Every other part of cortex not concerned with sensory processing or motor control.

atropine Cholinergic muscarinic receptor blocker.

auditory Associated with the sense of hearing.

autonomic nervous system (ANS) That part of the nervous system controlling our more vegetative functions.

autoreceptors Receptors found on neuron which respond to and modify release of its own transmitter.

axo-axonic synapse Where one axon terminal synapses on another. Important in mediation of presynaptic inhibition.

axodendritic synapse Where an axon terminal synapses onto dendrites.

axon The neuronal process which carries signals away from the cell body.

axon hillock Specialized region of membrane at start of axon.

axosomatic synapse Where an axon terminal synapses onto the cell soma.

ballistic movement Rapid pre-programmed movement.

baroreceptors Sensory receptors found in the aortic arch and carotid sinus sensitive to changes in blood pressure.

basket cell Neuronal type found at a number of sites in the CNS, the terminals of which cover the target cell rather like a woven basket and exert profound inhibition upon it.

Betz cell Largest of the pyramidal cells of the motorsensory cerebral cortex.

bilateral Concerning both sides of a structure.

blood–brain barrier Collective name for a variety of membranes and mechanisms which carefully control the substances permitted to reach the brain.

Brodmann's areas A map of the cerebral cortex in which anatomically defined areas are assigned numbers.

caudal Towards the tail.

centrifugal Impulse propagation away from higher centres of the nervous system.

centripetal Impulse propagation towards higher centres of the nervous system.

cephalization The gradual aggregation of neural material at the head end as a result of the evolutionary process.

cerebral dominance The extent to which a particular function is controlled by one side of the brain.

cerebrospinal fluid (CSF) Fluid surrounding brain and spinal cord.

characteristic frequency Input sound frequency to which auditory afferent is most sensitive.

choreiform movement Dance-like movement.

cingulate cortex Part of limbic system found on medial aspect of cerebral cortex.

clasp-knife reflex Sudden collapse of reflex muscle contraction after initial resistance to stretch. Seen in a decerebrate preparation, it is thought to be due to a change in gain of the interneurons mediating reflex inhibition from the Golgi tendon organs.

collateral A branch of an axon.

collateral sprouting Response from intact axons at partially denervated site.

condensation Region of maximum air pressure during sound-wave cycle.

cone Sensory receptor specialized for colour vision.

contralateral The opposite half of the body with respect to the point of reference.

convergence Imposition of information flow from many neurons onto a single one.

coronal Sections cut perpendicular to long axis of neuraxis, irrespective of bends. Thus in human a coronal section of forebrain is at right angles to one through spinal cord.

corpus callosum Fibre system running across brain connecting equivalent points on two sides.

co-transmission Two or more neuroactive substances are released from the same nerve terminal.

cranial nerves Twelve pairs of nerves arising from CNS at the level of the brainstem. Serve motor and sensory functions for head and some autonomic functions for viscera.

Dale's principle Knowledge of the transmitter released from a peripheral branch of a neuron suggests the nature of the transmitter employed at its central branches.

dark current Sodium-mediated current flow which results in the depolarization of the photoreceptor in the absence of stimulation.

decerebrate rigidity Increased tone in antigravity muscles leading to characteristic posture of extended limbs and tail, dorsiflexed head.

decussation Crossing of nerve pathways from one side of the brain to the other.

delayed auditory feedback Experiment to determine the significance of auditory feedback in speech control by introducing an artificial delay of 100–200 ms.

dendrite Short, branching process which carries signals towards the cell body.

dendrodendritic synapse Termination of a dendrite on the dendrites of another neuron.

denervation Loss of neural input.

denervation supersensitivity Loss of neural input to a region may cause proliferation of receptors on postsynaptic membrane. Individual becomes more sensitive to drugs which activate these receptors.

deoxyribonucleic acid (DNA) Information-storing molecule of the gene.

depolarization Movement of a transmembrane potential towards zero.

descending In a caudal direction towards lower parts of the nervous system.

dichotic listening Simultaneous presentation of different auditory messages to either ear.

disinhibition Removal of inhibition.

distal Further from the centre than point of reference.

divergence Flow of information from a single neuron onto many others.

dorsal At the back.

dorsal root ganglion Collection of cell bodies of sensory afferent fibres close to spinal cord.

efferent Carrying impulses away from some structure.

electroencephalography (EEG) Technique of recording electrical activity of the brain using electrodes attached to the scalp.

electromyography (EMG) Technique analagous to electrocardiography (ECG) but for skeletal muscle. Electrical activity associated with muscle contraction transmitted through electrically conductive body fluids and tissues and picked up by electrodes placed on overlying skin surface.

electrotonus Passive spread of current by the cable properties of the axon.

encephalization Evolutionary process leading to aggregation of neural material and function towards one end of the body.

end-plate potential (EPP) Localized, non-propagated, relatively long-lasting depolarization of end-plate region of muscle membrane resulting from the opening of large diameter, non-specialized ionic channels.

engram Pattern of preferred pathways thought to be the basis of memory.

enteroceptive (interoceptive) From inside the body.

equilibrium potential Potential difference which just opposes the tendency of ions to flow down their chemical concentration gradient in the opposite direction.

excitation secretion coupling Linkage between arrival of action potential at the nerve terminal and subsequent release of neurotransmitter.

excitatory postsynaptic potential (EPSP) Localized, non-propagated, relatively long-lasting depolarization of neuron membrane resulting from opening of sodium and potassium channels.

exteroceptive From outside the body.

extrafusal fibre Term applied to the muscle fibres outside the muscle spindle.

extrapyramidal (parapyramidal) system Components of the motor control system, the descending fibres of which do not pass through the medullary pyramids. (e.g. many of the descending influences on γ-motoneurons).

facilitation A small increase in neuronal excitability stopping short of overt discharge.

feedback Important control mechanism in which information from sensory receptors is relayed back to the CNS to modify subsequent output, e.g. in motor control and homeostasis.

fight or flight response Collection of biochemical, physiological and emotional responses activated by the sympathetic nervous system when an animal is in danger.

fissure A deep groove in the cerebral cortex.

frequency modulation Coding of changes in stimulus characteristics by frequency, pattern and number of action potentials produced.

frequency place code Separation of input sound frequencies along length of basilar membrane.

ganglion Aggregation of neuronal cell bodies outside the CNS.

gated channels (voltage-dependent channels) Ionic channels the opening of which is dependent on the membrane potential.

gene Length of DNA encoding information for a single protein.

generator current Current carried by ionic flow across membrane of receptor when permeability changed by stimulation.

generator potential Change in potential difference across membrane of receptor as a result of generator current.

glial cell Cells which perform supportive and metabolic functions for neurons.

Golgi tendon organ Tension-sensitive receptor found at junction between muscle fibres and tendon.

grey matter Region of brain containing a high percentage of neuronal cell bodies.

gustatory Associated with the sense of taste.

gyrus Convolution of the cerebral cortex.

hair cell Receptor cell of cochlea or vestibular apparatus.

homeostasis Maintenance of constancy of internal environment.

horizontal section Section parallel to back and front of neuraxis.

hyperpolarization Movement of a transmembrane potential further from zero.

idiopathic Of unknown cause.

impedance Resistance to an alternating waveform.

inferior colliculi Pair of nuclei in midbrain mediating auditory reflexes.

inhibitory postsynaptic potential Localized, non-propagated, relatively long-lasting hyperpolarization of neuron membrane as a result of opening of chloride channels.

initial segment Specialized region of axon next to axon hillock where action potentials arise.

interneuron A short-axoned neuron lying on the transmission path between two neurons or mediating feedback effects onto the same neuron.

intracellular Inside the cell.

intrafusal fibre Modified muscle fibre found inside the muscle spindle.

ionophore A molecule which promotes the passage of an ion across a membrane.

ipsilateral The same side of the body as the point of reference.

kinaesthesia Associated with the sense of movement.

lamina Region of concentration of neuronal type.

lateral Further from median plane (see medial) than reference structure.

laterality Assignment of particular function to one cerebral hemisphere.

masking Raising of audibility threshold of one frequency by introduction of second frequency.

medial Closer to the median plane (that plane which divides body into left and right halves) than reference structure.

microelectrode Probe made of conductive material (e.g. metal, carbon, glass filled with electrolyte) with a small recording surface (a few micrometres or less). Can be used to record from or stimulate single neurons.

mitochondria Cell organelle producing energy from aerobic respiration.

modality Qualitative nature of stimulus.

muscle spindle Length-sensitive receptor found in muscle.

myelin Fatty substance produced by glial cells which encases segments of some axons increasing their conduction velocity.

Nernst equation Describes the theoretical potential difference which would be produced across a membrane by diffusion of an ion down its concentration gradient if the membrane was freely permeable to that ion.

neuraxis The whole length of the central nervous system from forebrain to end of spinal cord.

neuromodulator Substance released from a neuron, often in addition to a neuro-transmitter, which may have a different, slower postsynaptic effect to the transmitter and may modify the response to that transmitter.

neuron Basic cellular information storage and transmission unit of nervous system.

neuropil Collective name for the nerve terminals and the dendrites with which they synapse.

neurotransmitter Substance released from neuron which carries information across the synapse and has an effect on the target cell, e.g. opening of ionic channels or release of second messengers. Actions of transmitters tend to be more rapid and shorter lasting than those of neuromodulators.

nociceptive Associated with the sense of pain.

node of Ranvier Exposed section of axon between adjacent segments of myelin at which action potential regenerates.

nucleus In neurophysiological terms describes a discrete group of histologically identifiable neurons in the CNS, usually having a common function.

nystagmus Small, constant, automatic eye movements.

olfactory Associated with the sense of smell.

ontogeny Embryological development of the individual.

opponent processing Pairs of colours having opposite effects on visual processing neurons in retina and lateral geniculate nucleus. Improves and adjusts colour contrast.

optic chiasm Place at which optic nerve fibres from nasal half of retinae cross to opposite side.

orthodromic Signal propagation in the normal direction, for instance impulse conduction in the axon away from the cell body.

Pacinian corpuscle Encapsulated, rapidly adapting mechanoreceptor.

Papez circuit Sequence of interconnected centres in limbic system important in memory. Includes hippocampus, mammillary bodies, parts of thalamus and cingulate gyrus of cortex.

parathyroid gland Small endocrine gland found in close anatomical proximity to the thyroid gland which is important in regulation of calcium and phosphate levels.

peptide Molecule formed from a short assembly of amino acids linked by peptide bonds.

perseveration Repetition of the same word.

phase-locking Tendency of neuronal discharge to coincide with a particular point on input wave leading to a greater regularity, though not necessarily any change in frequency, of discharge.

phasic A rapid reduction in output to a sustained input.

phylogenetic Concerning the evolutionary development of organisms.

pituitary gland Major endocrine gland which controls the functions of many other endocrine glands and is itself partially controlled by the hypothalamus.

placebo Favourable therapeutic effect achieved using an apparently inert agent.

plasticity Degree to which the CNS changes in response to experience or damage.

polymodal sensory receptors Sensory receptors responding to more than one cutaneous modality.

posterior At the back.

potentiation An increase in the magnitude of response to apparently identical stimuli.

presynaptic inhibition Long-lasting inhibition (tens or hundreds of milliseconds) caused by depolarization of presynaptic terminal with resulting reduction in amount of neurotransmitter released per action potential.

primary afferent depolarization Particular instance of presynaptic inhibition at terminals of sensory afferents.

proprioception Sensory input signalling body position, or position of bits of body relative to other bits.

proximal Closer to the centre than the reference point.

Purkinje cell Output neuron of cerebellum.

putative neurotransmitter A substance for which there is strong, but not conclusive, evidence of neurotransmitter action.

pyramidal cell Projection and output cell of cerebral cortex.

rarefaction Region of minimum pressure of sound wave cycle.

receptive field Physical domain over which receptor responds, e.g. area of skin for a cutaneous mechanoreceptor.

receptor A sensory receptor is a neuron or part of a neuron specialized to detect changes in a particular type of stimulus. At the pharmacological level a receptor is a molecular complex on the target cell membrane which combines with the neurotransmitter.

reflex A pattern of behaviour which is evoked consistently in response to a particular stimulus.

relative refractory period A short period of time following the absolute refractory period when the nerve axon is less excitable and a greater than normal stimulus is required to initiate an action potential.

Renshaw cell A short interneuron in the ventral horn of the spinal cord which mediates feedback inhibition onto the motoneuron pool.

resting membrane potential Potential difference which exists across a membrane in its unstimulated state. In the nerve axon it is close to the equilibrium potential for potassium ions.

retrograde degeneration Degeneration between site of axonal damage and cell body.

rod Photoreceptor specialized to respond to low light levels in the blue/green part of the visible spectrum.

rostral Towards the top end of the neuraxis.

sagittal section Vertical midline sections in the long axis of the CNS.

saltatory conduction Conduction in myelinated axon in which the action potential 'leaps' from node to node.

second messenger A substance released via a complex series of biochemical reactions following activation of a receptor by its neurotransmitter.

sodium–potassium pump An energy-consuming linked carrier system in the axonal membrane which maintains the relative ionic distributions by pumping potassium ions into the axon and sodium ions out.

spatial summation Adding together of two or more spatially separated events, e.g. EPSPs.

spectral colouring The peaks and notches imposed on the input sound frequencies by the anatomical features of the pinna of the ear.

stapedius reflex Reflex contraction of intrinsic muscles of middle ear triggered by loud sound leading to stiffening of ossicular chain and reduction in middle-ear sound conductance.

stimulus Something which provokes a response.

sulcus Shallow groove in the cerebral cortex.

superior colliculi Pair of nuclei in midbrain important in integration of some visual reflexes.

surround inhibition Phenomenon observed at many points in the nervous system where active neurons inhibit those around them in proportion to their level of activity. This eliminates low levels of activity and is used to improve spatial, colour and frequency contrast, etc.

synapse A structure for the transfer of information from one neuron to its target.

synaptic cleft A physical gap between the pre- and postsynaptic elements of a synapse.

synergist Muscle which acts in a similar fashion to the reference muscle.

tactile Concerning the sense of touch.

temporal summation Adding together of two or more temporally separated events.

threshold The point to which a neuron must be excited to produce an action potential.

tonic Applied to sensory receptors which show only a gradual reduction in output to a sustained input.

topography Mapping of area of brain with regard to some particular function, e.g. somatotopy.

tract Bundle of axons in the CNS.

transduction Conversion by a sensory receptor of the incident energy into electro-chemical energy.

trichromatic theory Theory, substantiated at the receptor level, that colour vision could be achieved by three sorts of colour receptors with the appropriate spectral sensitivities.

triple response Collective term for reflexes which facilitate focussing on a near object. These are accommodation, miosis and ocular convergence.

trophic substance A substance which promotes neuronal growth or development.

unilateral Concerning one side of a structure.

ventral At the front.

ventral horn Collection of grey matter in ventral part of spinal cord.

ventricles Series of interconnected cavities in brain filled with cerebrospinal fluid.

vesicle Membrane-bound neurotransmitter containing compartment in nerve terminal.

white matter Regions of CNS containing a high proportion of nerve fibres. Presence of myelin gives characteristic white colour.

Further reading and references

CHAPTER 1 — FUNCTIONAL NEUROANATOMY

Bowsher, D. (1979) *Introduction to the anatomy and physiology of the nervous system.* 4th ed., Blackwell, London.

Granit, R. (1967) *Receptors and sensory perception.* 5th edn., Yale University Press, New Haven, U.S.

CHAPTER 2 — NEURONAL FUNCTIONING

Aidley, D. J. (1978) *The physiology of excitable cells.* 2nd edn, Cambridge University Press, U.K.

Dale, H. H. (1935) *Pharmacology and nerve endings. Proc. Roy. Soc. Med.* **28** 319–332.

Fulton, B. P., Miledi, R. and Takahashi, T. (1980) Electrical synapses between motoneurons in the spinal cord of the newborn rat. *Proc. Roy. Soc. (Lond.) series B* **208** 115–120.

Katz, B. (1966) *Nerve, muscle and synapse.* McGraw-Hill, New York.

Keynes, R. D. and Aidley, D. J. (1981) *Nerve and muscle.* Cambridge University Press, U.K.

CHAPTER 3 — VISUAL SYSTEM

Davson, H. (1980) *Physiology of the eye.* 4th. edn, Churchill Livingstone, U.K.

Martin, K. A. C. (1988) From single cells to simple circuits in the cerebral cortex. *Quart. J. Exp. Physiol.* **73** (5) 637–702.

Stryker, M. P. (1989) Is grandmother an oscillation? *Nature* **338** 297–298.

Wiesel, N. T. and Gilbert, C. D. (1983) Morphological basis of visual cortical function. *Quart. J. Exp. Physiol.* **68** (4) 525–544.

CHAPTER 4 — AUDITORY SYSTEM

Holley, M. C. and Ashmore, J. F. (1988) A cytoskeletal spring in cochlear outer hair cells. *Nature* **335** 635–637.

Pickles, J. O. (1982) *An introduction to the physiology of hearing*. Academic Press, London.

von Békésy, G. (1960) *Experiments in hearing*. McGraw Hill, New York.

CHAPTER 5 — TACTILE SYSTEM

Iggo, A. (Ed.) (1973) *The somatosensory system*. Springer, Berlin.

Sinclair, D. (1981) *Mechanisms of cutaneous sensation*. Oxford University Press, U.K.

Wall, P. D. (1970) Sensory and motor role of dorsal columns. *Brain*, **93** 505–524.

CHAPTER 6 — TASTE AND SMELL

Moncrief, R. W. (1967) *The chemical senses*. Leonard Hill, London.

CHAPTER 7 — PAIN

Bonica, J. J. (Ed.) (1980) *Pain*. Raven Press, New York.

Bonica, J. J. and Albe Fessard, D. (Eds) (1976) *Advances in pain research and therapy, Vol. 1*, Raven Press, New York.

Brown, A. G. (1981) *Organization in the spinal cord*. Springer, Berlin.

Davidoff, R. A. (Ed.) (1984) *Handbook of the spinal cord*. Decker, New York.

Holden, A. V. and Winlow, W. (Ed.) (1985) *The neurobiology of pain*. Manchester University Press, U.K.

Hughes, J. (1985) Isolation of an endogenous compound from the brain with properties similar to morphine. *Brain Research*, **88** 295–308.

Iggo, A. and Kornhuber, H. H. (1975) A quantitative study of C-mechanoreceptors in hairy skin of the cat. *J. Phys.* **271** 549–565.

Keele, K. D. (1957) Anatomies of pain. Oxford University Press.

Melzack, R. (1973) *The puzzle of pain*. Penguin, London, U.K.

Melzack, R. and Wall, P. D. (1965) Pain mechanisms: a new theory. *Science*, 971–979.

Ogden, T. E., Robert, F. and Carmichael, E. A. (1959) Some sensory syndromes in children: Indifference to pain and sensory neuropathy. *J. Neurol. Neurosurg. Psychiat.* **22** 267–276.

Roberts, M. H. T. (1984) 5-Hydroxytryptamine and nociception. *Neuropharmacology* **23** (12B) 1529–1536.

CHAPTER 8 — PROPRIOCEPTION AND KINAESTHESIA

Wilson, V. J. and Melvill Jones, G. (1979) *Mammalian vestibular physiology*. Plenum Press, New York.

See also references in Chapter 9

144   Further reading and references

CHAPTER 9 MOTOR CONTROL

Asanuma, H. (1975) Recent developments in the study of the columnar arrangements of neurons within the motor cortex. *Physiol. Rev.* **55** (2), 143–156.

Barnes, C. D. and Schadt, J. C. (1979) Release of function in the spinal cord. *Prog Neurobiol.* **12** (1) 1–15.

Boyd, I. A. and Gladden, M. H. (Eds) (1985) *The muscle spindle.* Macmillan Press, U.K.

Davson, H. and Segal, M. B. (1978) *Introduction to Physiology,* Vol. 4, Academic Press, London.

Dray, A. (1980) The physiology and pharmacology of mammalian basal ganglia. *Prog. Neurobiol.* **14**, 221–335.

Eccles, J. C. (1973) *The understanding of the brain.* McGraw-Hill, New York.

Edgley, S. A.and Lidierth, M. (1987) The discharges of cerebellar Golgi cells during locomotion in the cat. *J. Physiol.* **392** 315–332.

Garcia-Rill, E. (1986) The basal ganglia and the locomotor regions. *Brain Research Reviews* **11** 47– 63.

Hasan, Z. and Stuart, D.G. (1988) Animal solutions to problems of movement control: the role of proprioceptors. *Ann. Rev. Neurosci.* **11** 199–223.

Krishna Murthy, K. S. (1978) Vertebrate fusimotor neurones and their influences on motor behaviour. Prog. Neurobiol. **11** 249–307.

Matthews, P. B. C. (1977) Muscle afferents and kinaesthesia. *British Medical Bulletin* **33** (3) 137.

Matthews, P. B. C. (1981) Evolving views on the internal operation and functional role of the muscle spindle. *J. Physiol.* **320**, 1–30.

Phillips, C. G. and Porter, R. (1977) *Corticospinal neurones: their role in movement.* Academic Press, London.

Porter, R. (1985) The corticomotoneuronal component of the pyramidal tract: corticomotoneuronal connections and functions in primates. *Br. Res. Reviews* **10** (1) 1–27.

Sacks, O. (1982) *Awakenings.* Pan Books Ltd., London.

Shahani, M. (Ed.) (1976) The motor system, neurophysiology and muscle mechanisms. *Satellite Symposium, 26th International Congress of Physiology, India 1974.* Elsevier Ltd., Amsterdam.

Taylor, A. and Prochazka, A. (Eds.) (1981) *Muscle receptors and movement.* Macmillan, London.

CHAPTER 10 — AUTONOMIC NERVOUS SYSTEM

Iversen, L. L., Iversen, S. D. and Snyder, S. H. (Eds) (1978) *Handbook of psychopharmacology.* Plenum Press, New York.

Mountcastle, V. B (Ed.) (1968) *Medical physiology, Vol 2,* C. V. Mosby, St. Louis.

Tucek, S. (Ed.) (1979) The cholinergic synapse. *Prog. Br. Res.* **49**, Elsevier, New York.

CHAPTER 11 — CORTICAL LOCALIZATION, LATERALIZATION AND SPEECH

Borden, G. J. and Harris, K. S. (1980) Speech science primer. Physiology, acoustics and perception of speech. Williams and Wilkins, Baltimore.

Espir, M. L. E. and Rose, F. C. (1976) *The basic neurology of speech.* 2nd ed., Blackwell, Oxford.

Geschwind, N. (1974) *Language and the brain.* Reidel, Doordrecht, Netherlands.

Kolb, B. and Whishaw, I. Q. (1985) *Fundamentals of human neuropsychology.* 2nd edn. Freeman, New York.

Peterson, S. E., Fox, P. T., Posner, M. I., Mintum, M. and Raichle, M. E. (1988) Positron emission tomographic studies of the cortical anatomy of single word processing. *Nature* **331** 585–589.

Zangwill, O. L. (1960) *Cerebral dominance and its relation to psychological function.* Charles, C. Thomas. Springfield, Illinois.

CHAPTER 12 — PLASTICITY

Bjorkland, A., Segal, M. and Stenevi, U. (1979) Functional reinnervation of rat hippocampus by locus coeruleus implants. *Br. Res.* **170** 409–426.

Cajal, S. R., (1928) *Degeneration and regeneration of the nervous system.* Oxford University Press.

Horridge, G. A. (1965) The electrophysiological approach to learning in isolatable ganglia. *Animal Behaviour,* Supplement 1 163–182.

Sperry, R. W. (1965) Mechanisms of neural maturation. In: Stevens, S. S. (Ed.) *Handbook of experimental neurology.* Wiley, New York.

Stein, D. G., Rosen, J. J. and Butters, N. (Eds.) (1974) *Plasticity and recovery of function in the central nervous system.* Academic Press, New York.

Rosenzwieg, M. R. Bennett, E. L. and Diamond, M. L. (1972) Brain changes in response to experience. *Scientific American,* **226** 22–29.

Wall, P. D. and Egger, M. D. (1971) Formation of new connections in adult rat brains after partial deafferentiation. *Nature,* **232** 542–545.

Waller, A. V. (1850) Experiments on the section of the glossopharyngeal and hypoglossal nerves of the frog, and observations of the alterations produced thereby in the structure of their primitive fibres. *Philosophical Transactions,* **140** 423–429.

CHAPTER 13 — NEUROCHEMICAL ASPECTS OF MENTAL FUNCTION

Bradford, H. F. (1986) *Chemical neurobiology: an introduction to neurochemistry.* Freeman. New York.

Cooper, J. R., Bloom, F. E. and Roth, R. H. (1982) *The biochemical basis of neuropharmacology.* 4th. edn. Oxford University Press.

Feldman, R. S. and Quenzer, L. F. (1984) *Fundamentals of neuropsychopharmacology.* Sinauer Associates, Sunderland, Massachusetts.

Fluckiger, E., Muller, E. E. and Thorner, M. O. (Eds) (1987) Transmitter molecules in the brain. *Basic and clinical aspects of neuroscience, Vol. 2,* 1–78.

Iversen, S. D. and Iversen, L. H. (1977) *Behavioural Pharmacology.* Oxford University Press.

Kruk, Z. L. and Pycock, C. J. (1983) *Neurotransmitters and drugs.* 2nd. edn., Croom Helm,. Australia.

Loewi, O. (1921) Ueber humorale Uebertragbarkeit der Herznervenwirking. *Pflugers Archiv.* **189** 239–242.

Usdin, E., Hamburg, D. A. and Barchas, J. D. (Eds) (1977) *Neuroregulators and psychiatric disorders.* Oxford University Press.

Index